W9-BRX-192

Laboratory Manual for Physical Geology
Eighth Edition

James H. Zumberge

Professor of Geosciences and President,
University of Southern California

Robert H. Rutford

Professor of Geology and President,
University of Texas at Dallas

 WCB Wm. C. Brown Publishers

Book Team

Editor *Jeffrey L. Hahn*
Developmental Editor *Lynne M. Meyers*
Production Editor *Diane Clemens*
Designer *Laurie J. Entringer*
Art Editor *Barbara J. Grantham*
Photo Editor *Carrie Burger*
Visuals Processor *Kenneth E. Ley*
Permissions Editor *Mavis Oeth*

WCB **Wm. C. Brown Publishers**

President *G. Franklin Lewis*
Vice President, Publisher *George Wm. Bergquist*
Vice President, Publisher *Thomas E. Doran*
Vice President, Operations and Production *Beverly Kolz*
National Sales Manager *Virginia S. Moffat*
Advertising Manager *Ann M. Knepper*
Marketing Manager *John W. Calhoun*
Editor in Chief *Edward G. Jaffe*
Managing Editor, Production *Colleen A. Yonda*
Production Editorial Manager *Julie A. Kennedy*
Production Editorial Manager *Ann Fuerste*
Publishing Services Manager *Karen J. Slaght*
Manager of Visuals and Design *Faye M. Schilling*

Cover photo by © David Muench 1990

Photos: p. v (bottom) NASA ERTS E-2272-14543, U.S.G.S. EROS
Data Center, Sioux Falls, South Dakota 57198; p. vi Courtesy of NASA
and the U.S.G.S. EROS Data Center, Sioux Falls, South Dakota 57198.

Figures 1.1, 1.2, 1.5–1.20, 1.25–1.40, 1.43–1.47, 1.49, 1.50, 1.53–1.61:
© Wm. C. Brown Publishers, Robert Rutford, and James Zumberge/
James Carter, photographer.

Copyright © 1951, 1957, 1967, 1973, 1979, 1983, 1988, 1991 by James
H. Zumberge. All rights reserved

Library of Congress Catalog Card Number: 90–81910

ISBN 0-697-09844-3

No part of this publication may be reproduced, stored in a retrieval
system, or transmitted, in any form or by any means, electronic,
mechanical, photocopying, recording, or otherwise, without the
prior written permission of the publisher.

Printed in the United States of America by Wm. C. Brown Publishers,
2460 Kerper Boulevard, Dubuque, IA 52001

10 9 8 7 6 5 4 3 2 1

Contents

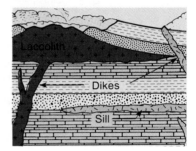

Aerial Photographs and Other Imagery from Remote Sensing   68

Part 3   Geologic Interpretation of Topographic Maps, Aerial Photographs, and Earth Satellite Images   77

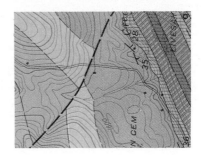

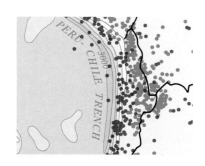

Materials Needed by Students Using This Manual

1. Scale ("ruler") graduated in tenths of an inch.
2. Colored pencils (red, blue, and assorted other colors).
3. Felt tip pens ($\frac{1}{8}'' \times \frac{1}{4}''$ tip), three assorted colors.
4. Several medium to medium-soft pencils (2H or No. 2).
5. Pocket stereoscope.
6. Small magnifying glass (optional) for map reading.
7. Six sheets $8\frac{1}{2}'' \times 11''$ tracing paper.
8. Eraser (art gum or equivalent).
9. Inexpensive pencil sharpener.
10. Inexpensive compass, for drawing circles.
11. Dividers (optional), for measuring distances on maps.

Preface

The geological sciences have undergone remarkable changes since the first edition of this manual was published in 1951. Those changes that have endured over time have been incorporated in each edition of this manual published since mid-century. Although the subject matter has changed and expanded in scope, the number of laboratory sessions in a given academic quarter or semester has not increased. Because the time available in a semester or quarter cannot be expanded without disrupting the class schedule for the entire college or university, the problem of too much subject material for too little time poses a dilemma for both authors and instructors.

On the assumption that the subject material to be covered in any course is the prerogative of the instructor and not the authors, we have written a manual that contains more material than can be covered in a single laboratory course, thereby leaving the selection of individual exercises to the instructor. While we believe that the overall scope of this manual is in keeping with the general subject material covered in a beginning laboratory course, we think that instructors should determine the specific exercises that are in keeping with their own ideas of how to organize and present the subject material.

In addition to the variety of laboratory exercises offered, we have also expanded portions of the background material that accompanies each exercise. By allowing students to review the important geologic terms and concepts they will encounter in the laboratory, we hope to enhance their chances for successful completion of the exercises. According to reviewers and users of the previous edition, this supplementary information should be particularly useful in those instances where students do not routinely bring their textbooks to the lab or where students are not concurrently enrolled in the lecture course.

The eighth edition follows the same overall organization used in the seventh edition. Part 1, "Earth Materials," has been expanded in response to suggestions from users. To ease the difficulties that students sometimes encounter in identifying unknown rocks and minerals, the identification keys have been revised, and all new rock and mineral hand samples have been photographed expressly for this edition.

Part 2, "Topographic Maps, Aerial Photographs, and Other Imagery from Remote Sensing," contains a few refinements but remains otherwise unchanged.

Part 3, "Geologic Interpretation of Topographic Maps, Aerial Photographs, and Earth Satellite Images," has been modified by the addition of new exercises, deletion of older exercises, and the inclusion of background explanatory materials. Also, many of the exercises have been shortened.

Part 4, "Structural Geology," follows the traditional approach, but we have added material on the relationship between faults and earthquakes and have added a new exercise on the location of an epicenter.

Part 5, "Plate Tectonics and Related Geologic Phenomena," has an expanded section on the relationship of magnetic anomalies to the rates of seafloor spreading along the East Pacific Rise and the Mid-Atlantic Ridge. There is also a new exercise on the reconstruction of the relative positions of the coastlines of Africa and South America at a given time in the geologic past.

We are especially grateful to Charles G. Sammis of the University of Southern California and Peter J. Wyllie of the California Institute of Technology for their help in preparing the materials on earthquakes and plate boundaries, respectively. James Carter of the University of Texas at Dallas took all of the new mineral and rock photographs included in Part 1. Diagrams and figures from *Fundamentals of Geology* (2nd edition), by Carla W. Montgomery (1990, Wm. C. Brown Publishers), and *Physical Geology* (4th edition), by Charles C. Plummer and David McGeary (1988, Wm. C. Brown Publishers), are acknowledged where they occur.

To those who reviewed this edition, we express our thanks and appreciation for their critical comments and suggestions for improvement. These include Dr. Mary Jo

Richardson, Texas A & M University; Professor Vicki Harder, Texas A & M University; Professor Roseann J. Carlson, Tidewater Community College; Professor Harold Stowell, University of Alabama; and Professor Ray Kenny, Arizona State University.

As authors, we accept the full responsibility for any inadvertent errors that have crept into these pages, and we welcome comments from users if they discover such errors.

Finally, to the professional men and women of Wm. C. Brown Publishers, we extend our gratitude for their design of the format and expert editorial help in transforming our manuscript into this finished product.

James H. Zumberge
University of Southern California

Robert H. Rutford
University of Texas at Dallas

Earth Materials

Background

The materials that make up the crust of the earth fall into two broad categories: minerals and rocks. Minerals are elements or chemical compounds that are formed by a number of natural processes. Rocks are aggregates of minerals or organic substances that occur in many different architectural forms over the face of the earth, and they contain a significant part of the geologic history of the region where they occur. To identify them and understand their history, it is necessary to be able to identify the minerals that make up the rocks.

The first goal of Part 1 is to introduce beginning students of geology to the identification of minerals and rocks through the use of simplified identification methods and classification schemes. Students will be provided with samples of minerals and rocks in the laboratory. These samples are called *hand specimens*. Ordinarily their study does not require a microscope or any means of magnification because the naked or corrected eye is sufficient to perceive their diagnostic characteristics. A feature of a mineral or rock that can be distinguished without the aid of magnification is said to be *macroscopic* (also *megascopic*) in size. Conversely, a feature that can be identified only with the aid of magnifiers is said to be *microscopic* in size. The exercises that deal with the identification and classification of minerals and rocks in Part 1 are based only on macroscopic features.

The second goal of this part is to acquaint the student with the modes of formation and occurrence of rocks and their relative age relationships. Many students using this manual will participate in organized field trips to observe first-hand how rocks occur in nature. To prepare students for this field experience, some basic geological principles on the mode of occurrence of rocks will be introduced in the form of simple geologic diagrams. The concept of geologic time will be explored, and the relative age relationships of rock masses will be examined according to some basic geologic principles and assumptions.

Even if organized field trips are not a required part of your geology course, an understanding of these geologic diagrams will enhance your appreciation of rock strata and other geologic phenomena as you encounter them in your travels.

Minerals

Definition

A mineral is a naturally occurring, crystalline, inorganic, homogeneous solid with a chemical composition that is either fixed or varies within certain fixed limits, and a characteristic internal structure manifested in its exterior form and physical properties.

Mineral Identification

Common minerals are identified or recognized by testing them for general or specific physical properties. For example, the common substance table salt is actually a mineral composed of sodium chloride (NaCl) and bears the mineral name halite. The taste of halite is distinctive and is sufficient for identifying it and distinguishing it from other substances such as sugar (not a mineral) that have a similar appearance. Chemical composition alone is not sufficient to identify minerals. The mineral graphite and the mineral diamond are both composed of a single element, carbon (C), but their physical properties are very different and distinct.

The taste test applied to halite is restrictive because it is the only mineral that can be so identified. Other minerals may have a specific taste, but it is different from that of halite. Other common minerals can be tested by visual inspection for the physical properties of crystal form, cleavage, or color or by using simple tools such as a knife blade or glass plate to test for the physical property of hardness.

The first step in learning how to identify common minerals is to become acquainted with the various physical properties that individually or collectively characterize a mineral specimen. These are described in the following paragraphs.

Properties of Minerals

The physical properties of minerals are those that can be observed generally in all minerals. They include such common features as luster, color, hardness, cleavage, streak, and specific gravity. *Other properties* are those that are found in only a few minerals. These include taste, odor, magnetism, and chemical reaction with acid. In your work in the laboratory, use the hand specimens sparingly when applying tests for the various properties.

General Physical Properties

Luster

The appearance of a fresh mineral surface in reflected light is its luster. A mineral that looks like a metal is said to have a *metallic luster*. Minerals that are *non-metallic* are described by one of the following adjectives: *vitreous* (having the luster of glass); *resinous* (having the luster of resin); *pearly; silky; dull* or *earthy* (not bright or shiny).

Your laboratory instructor will display examples of minerals that possess these various lusters.

Color

The color of a mineral is determined by examining a fresh surface in reflected light. Color and luster are not the same. Some minerals are clear and transparent, thus colorless. The color or lack of color may be diagnostic in some minerals, but in others, the color varies due either to a slight difference in chemical composition or to small amounts of impurities within the mineral (fig. 1.1).

Hardness

The hardness of a mineral is its resistance to abrasion. Hardness can be determined either by trying to scratch a mineral of unknown hardness with a substance of known hardness, or by using the unknown mineral to scratch a substance of known hardness. Hardness is measured on a relative scale called the *Mohs scale of hardness*,[1] which consists of ten common minerals arranged in order of their increasing hardness (table 1.1). In the laboratory, convenient materials other than these ten specific minerals may be used for hardness determination.

In this manual, a mineral that scratches glass will be considered "hard," and one that does not scratch glass will be "soft." In making hardness tests on a glass plate, do not hold the glass in your hands; keep it firmly on the table

---

1. Friedrich Mohs (1773–1839) was a German mineralogist and protégé of the famous geologist-mineralogist, Abraham Gottlob Werner (1750–1817) of Freiburg, Germany.

Table 1.1 Mineral Hardness According to the Mohs Scale (*A*) and Some Common Materials (*B*)

HARDNESS	A	B
1	TALC	
2	GYPSUM	
2.5		FINGERNAIL
3	CALCITE	
3.5		COPPER PENNY
4	FLUORITE	
5	APATITE	
5–5.5		KNIFE BLADE
5.5		GLASS PLATE
6	ORTHOCLASE	
6.5		STEEL FILE
7	QUARTZ	
8	TOPAZ	
9	CORUNDUM	
10	DIAMOND	

top. If you think that you have made a scratch on the glass, try to rub the scratch off. What appears to be a scratch may be only some of the mineral that has rubbed off on the glass.

Cleavage

Cleavage is the tendency of some mineral crystals to break along definite planes of weakness that exist in the internal (atomic) structure of the crystal. Cleavage planes are related to the crystal system of the mineral and are always parallel to crystal faces or possible crystal faces. Cleavage may be conspicuous and is a characteristic physical property that is useful in mineral identification. In some minerals it is almost impossible to break a crystal in such a way that cleavage planes do not develop. An example of this is the characteristic rhombohedral cleavage of calcite.

Perfect cleavage describes cleavage planes that are very smooth and flat and that reflect light much like a mirror. Other descriptors such as *good, fair,* and *poor* are used to describe cleavage surfaces that are less well defined. Some minerals exhibit excellent crystal faces but have no cleavage; quartz is such a mineral.

The cleavage planes of some minerals such as calcite, muscovite (fig. 1.2), halite, and fluorite are so well developed that they are easily detected. In others, the cleavage surfaces may be so discontinuous as to escape detection by casual inspection. Before deciding that a mineral has no cleavage, turn it around in a strong light and observe whether there is some position in which the surface of the specimen "lights up," that is, reflects the light as if it were the reflecting surface of a dull mirror. If so, the mineral has cleavage, but the cleavage surface consists of several discontinuous parallel planes minutely separated, and rather than perfect cleavage it has good, fair, or poor cleavage. As noted in the discussion of crystal form, it is important to differentiate between cleavage planes and crystal faces; the actual breaking of a mineral crystal may be useful in making this differentiation.

Figure 1.1 The specimens in this photograph are all quartz. The difference in colors is due to various impurities. Clockwise from left: smoky quartz, quartz crystal, rose quartz, citrine quartz, amethyst quartz.

Figure 1.2 Muscovite is a mineral with excellent basal cleavage.

In assigning the number of cleavage planes to a specimen, do not make the mistake of calling two parallel planes bounding the opposite sides of a specimen two cleavage planes. In this case, the specimen has two cleavage surfaces but only one plane of cleavage (i.e., one direction of cleavage). For example, halite has cubic cleavage, thus six sides, but only three planes of cleavage, because the six sides are made up of three parallel pairs of cleavage planes.

The angle at which two cleavage planes intersect is diagnostic. This angle can be determined by inspection. In most cases you will need to know whether the angle is 90 degrees, almost 90 degrees, or more or less than 90 degrees. The cleavage relationships that you will encounter during the course of your study of common minerals are tabulated for convenience in figure 1.3.

Number of Cleavage Planes	Remarks	Examples
1	Usually called *basal* cleavage. **Muscovite and** biotite are examples.	
2	Two at 90 degrees. Feldspar and pyroxene (augite) have cleavage surfaces that intersect at close to 90 degrees.	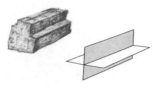
2	Two *not* at 90 degrees. Amphibole (hornblende) cleavage surfaces intersect at angles of about 60 and 120 degrees.	
3	**Three at 90 degrees. Minerals with three planes of cleavage that intersect at 90 degrees are said to have *cubic* cleavage.** Halite and galena are examples.	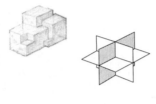
3	**Three *not* at 90 degrees. A mineral which breaks into a six-sided prism, with each side having the shape of a parallelogram, has *rhombic* cleavage.** Example: calcite.	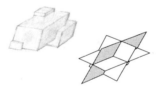
4	**Four sets of cleavage surfaces in the form of an octahedron produce *octahedral* cleavage.** Example: fluorite.	
6	**Complex geometric forms.** Example: sphalerite.	

Figure 1.3 Descriptive notes on cleavage planes. (Adapted and reprinted with permission from R. D. Dallmeyer, *Physical Geology Laboratory Manual*, Dubuque, Iowa. Kendall/Hunt Publishing Company. Copyright © 1978 by Kendall/Hunt Publishing Company.)

Some minerals exhibit the characteristic of *parting*, sometimes called false cleavage. Parting occurs along planes of weakness in the mineral, but usually the planes are more widely separated and often are due to twinning deformation or inclusions. Parting is not present in all specimens of a given mineral.

Fracture generally refers to breakage that forms a surface with no relationship to the internal structure of the mineral crystal; that is, the break occurs in a direction other than a cleavage plane. Quartz is characterized by *conchoidal* fracture (the fracture surfaces are smooth and exhibit fine concentric ridges), asbestos by *fibrous* fracture. Other terms often used include *hackly, uneven* (rough), *even* (smooth), and *earthy* (dull but smooth fracture surfaces, common in soft mineral aggregates such as kaolinite).

Streak

The color of a mineral's powder is its streak. The streak is determined by rubbing the hand specimen on a piece of unglazed porcelain (*streak plate*). Some minerals have a streak that is the same as the color of the hand specimen; others have a streak that differs in color from the hand specimen. The streak of minerals with a metallic luster is especially diagnostic.

Specific Gravity

The specific gravity (G) of a mineral is a number that represents the ratio of the mineral's weight to the weight of an equal volume of water.[2] In contrast to density, defined as weight per unit volume, specific gravity is a dimensionless number. The higher the specific gravity, the greater the density of a mineral.

For purposes of determining specific gravity of the minerals in the laboratory, it is sufficient to utilize a simple "heft" test by lifting the hand specimens. Compare the heft of two specimens of about the same size but of contrasting specific gravity. For example, heft a specimen of graphite (G = 2.2) in one hand while hefting a similar sized specimen of galena (G = 7.6) in the other hand. *Take care to compare only similar sized samples.* This test allows you to determine the relative specific gravity of minerals. Minerals such as graphite (G = 2.2) and gypsum (G = 2.3) are relatively light when hefted. Quartz (G = 2.6) and calcite (G = 2.7) are of average heft, whereas corundum (G = 4.0), magnetite (G = 5.2), and galena (G = 7.6) are heavy when hefted.

Diaphaneity

The ability of a thin slice of a mineral to transmit light is its diaphaneity. If a mineral transmits light freely so that an object viewed through it is clearly outlined, the mineral

2. A detailed description of the method of determining specific gravity in the laboratory is given in *Minerals and How to Identify Them*, by E. S. Dana, 3d ed., revised by C. S. Hurlbut, Jr. (New York: John Wiley and Sons, Inc., 1963), pp. 75–80.

is said to be *transparent*. If light passes through the mineral but the object viewed is not clearly outlined, the mineral is *translucent*. Some minerals are transparent in thin slices and translucent in thicker sections. If a mineral allows no light to pass through it, even in the thinnest slices, it is said to be *opaque*.

Tenacity

This property is an index of a mineral's resistance to being broken or bent. It is not to be confused with hardness. Some of the terms used to describe tenacity are:

Brittle—The mineral shatters when struck with a hammer or dropped on a hard surface.

Elastic—The mineral bends without breaking and returns to the original shape when stress is released.

Flexible—The mineral bends without breaking but does not return to its original shape when the stress is released.

Crystal Form

A crystal is a solid bounded by smooth surfaces (crystal faces) that reflect the internal (atomic) structure of the mineral. *Crystal form* refers to the assemblage of faces that constitute the exterior surface of the crystal. *Crystal symmetry* is the geometric relationship between the faces.

Seven crystal systems are recognized by crystallographers, and all crystalline substances crystallize in one of the seven crystal systems (fig. 1.4). Some common substances, such as glass, are often described as crystalline, but in reality they are *amorphous;* that is, they have solidified with no fixed or regular internal atomic structure.

The same mineral always shows the same angular relations between crystal faces, a relationship known as the *constancy of interfacial angles*. The symmetric relationship of crystal faces, related to the constancy of interfacial angles, is the basis for the recognition of the crystal systems.

Symmetry in a crystal is determined by completing a few geometric operations. For example, a cube has six faces, each at right angles to the adjacent faces. A planar surface that divides the cube into portions such that the faces on one side of the plane are mirror images of the faces on the other side of the plane is called a *plane of symmetry*. A cube has nine such planes of symmetry. In the same way, imagine a line (axis) connecting the center of one face on a cube with the center of the face opposite it. Rotation of the cube about this axis will show that during a complete rotation a crystal face identical with the first face observed will appear in the same position four times. This is a *four-fold axis of symmetry*. Rotation of the cube around an axis connecting opposite corners will show that three times during a complete rotation an identical face appears, thus a *three-fold axis of symmetry*.

The seven crystal systems can be recognized by the symmetry they display. Figure 1.4 summarizes the basic elements of the symmetry for each system and shows some examples of the *crystal habit* (the crystal form commonly taken by a given mineral) of some minerals you may see in the laboratory or a museum.

Perfect crystals in nature are the exception rather than the rule. They usually form under special conditions where there is open space for them to grow during crystallization. Crystals more commonly are small and distorted. Nevertheless, the internal arrangement of the atoms is fixed, although the external form is not perfectly developed.

Many of the hand specimens you see in the laboratory will be made up of many minute crystals so that few crystal faces, or none, can be seen, and the specimen will appear granular. Other hand specimens may be fragments of larger crystals so that only one or two imperfect crystal faces can be recognized. Although perfect crystals are rare, most student laboratory collections contain some reasonably good crystals of quartz, calcite, gypsum, fluorite, and pyrite.

Two or more crystals of some minerals may be grown together in such a way that the individual parts are related through their internal structures. The external form that results is manifested in a *twinned crystal*. Some twins appear to have grown side-by-side (plagioclase), some are reversed or are mirror images (calcite), and others appear to have penetrated one another (fluorite, orthoclase, staurolite). Recognition of twinned crystals may be useful in mineral identification.

A word of caution: Cleavage fragments of minerals such as halite, calcite, and gypsum are often mistaken for crystals. This error is made because the cleavage fragments of these minerals have the same geometric form as the crystal.

Special Properties

Magnetism

The test for magnetism requires the use of a common magnet or magnetized knife blade. Usually, magnetite is the only mineral in your collection that will be attracted by a magnet.

Double Refraciton

If an object appears to be double when viewed through a transparent mineral, the mineral is said to have double refraction. Calcite is the best common example.

Taste

The saline taste of halite is an easy means of identifying the mineral. Few minerals are soluble enough to possess this property. (For obvious sanitary reasons, do not use the taste test on your laboratory hand specimens.)

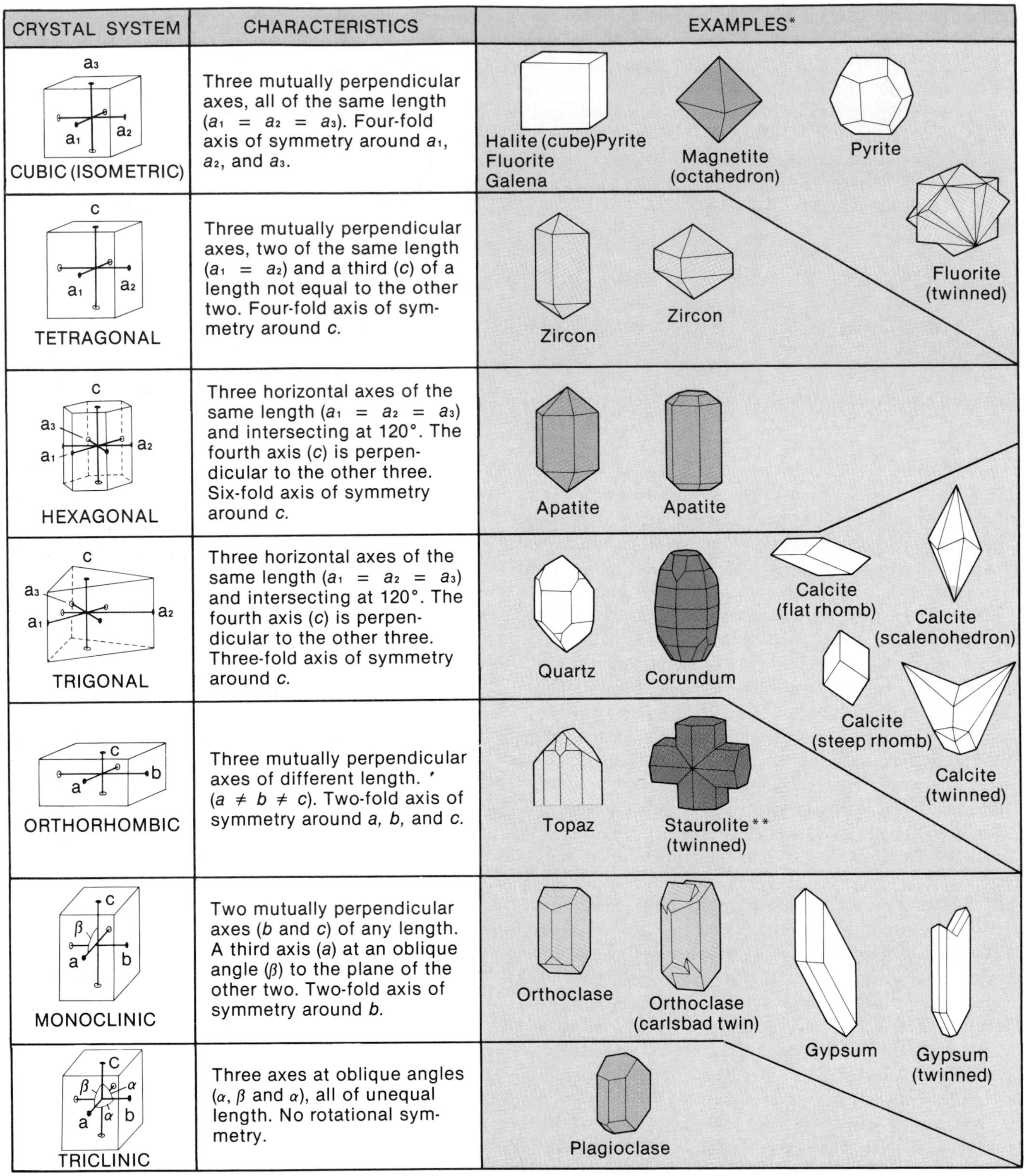

CRYSTAL SYSTEM	CHARACTERISTICS	EXAMPLES*
CUBIC (ISOMETRIC)	Three mutually perpendicular axes, all of the same length ($a_1 = a_2 = a_3$). Four-fold axis of symmetry around a_1, a_2, and a_3.	Halite (cube) Pyrite Fluorite Galena Magnetite (octahedron) Pyrite Fluorite (twinned)
TETRAGONAL	Three mutually perpendicular axes, two of the same length ($a_1 = a_2$) and a third (c) of a length not equal to the other two. Four-fold axis of symmetry around c.	Zircon Zircon
HEXAGONAL	Three horizontal axes of the same length ($a_1 = a_2 = a_3$) and intersecting at 120°. The fourth axis (c) is perpendicular to the other three. Six-fold axis of symmetry around c.	Apatite Apatite
TRIGONAL	Three horizontal axes of the same length ($a_1 = a_2 = a_3$) and intersecting at 120°. The fourth axis (c) is perpendicular to the other three. Three-fold axis of symmetry around c.	Quartz Corundum Calcite (flat rhomb) Calcite (scalenohedron) Calcite (steep rhomb) Calcite (twinned)
ORTHORHOMBIC	Three mutually perpendicular axes of different length. ($a \neq b \neq c$). Two-fold axis of symmetry around a, b, and c.	Topaz Staurolite** (twinned)
MONOCLINIC	Two mutually perpendicular axes (b and c) of any length. A third axis (a) at an oblique angle (β) to the plane of the other two. Two-fold axis of symmetry around b.	Orthoclase Orthoclase (carlsbad twin) Gypsum Gypsum (twinned)
TRICLINIC	Three axes at oblique angles (α, β and α), all of unequal length. No rotational symmetry.	Plagioclase

Figure 1.4 Characteristics of the seven crystal systems and some examples. (Copyright © 1974 McGraw-Hill Book Co. (UK) Ltd. From Cox, Price & Harte: *An Introduction to the Practical Study of Crystals, Minerals and Rocks,* Revised Edition. Reproduced by permission. Colors have been added to the original and are not accurate. They are shown for illustrative purposes only.) *Most laboratory collections of minerals for individual student use do not include crystals of these minerals. The collection may, however, contain incomplete single crystals, fragments of single crystals, or aggregates of crystals of one or more minerals. The best examples of these and other crystals may be seen on display in most mineralogical museums. **Staurolite is actually monoclinic but is also classified as pseudo-orthorhombic. Pseudo-orthohombic means that staurolite *appears* to be orthorhombic because the angle β in the monoclinic system (see left-hand column under monoclinic) is so close to 90 degrees that in hand specimen it is not possible to discern that the angle β for staurolite is actually 89 degrees, 57 minutes.

Exercise 1. Identification of Common Minerals

Your instructor will provide you with a variety of minerals to be identified. Take time to examine the minerals and review the various physical properties described in the previous pages. Select several samples and examine them for luster, color, hardness, and streak, and compare their specific gravity (G) using the heft test.

Study table 1.2. Note that certain minerals have physical properties that make their identification relatively easy. For example, graphite is soft, feels greasy, and marks both your hands and paper. Galena is "heavy," shiny, and has perfect cubic cleavage. Calcite has perfect rhombohedral cleavage, is easily scratched by a knife, reacts with cold dilute HCl, and in clear specimens shows double refraction.

When you feel that you have an understanding of the various physical properties and the tests that you must apply to determine these properties, select a specimen at random from the group of minerals provided to you in the laboratory. Follow the steps outlined below. Refer to figures 1.5 through 1.20 as an aid to identification. Be aware that your laboratory collection may contain some minerals that are not shown in the figures. Due to the normal variations within a single mineral species, some of the specimens may appear different from the same minerals shown in the figures in this manual.

1. Examine carefully a single mineral specimen selected at random from the group provided to you in the laboratory.

2. Determine whether the sample has a metallic or non-metallic luster. Then determine whether it is light- or dark-colored. (The terms *dark* and *light* are subjective. A mineral that is "dark" to one observer may be "light" to another. This possibility is anticipated in table 1.2 where mineral specimens that could fall into either the "light" or "dark" categories are listed in both groups. The same is true for minerals that may exhibit either metallic or non-metallic luster.)

3. If the mineral falls into either group I or II, proceed to test it first for hardness and then for cleavage. This will place the mineral with a small group of other minerals in table 1.2. Identification can be completed by noting other diagnostic general or specific physical properties.

   If the mineral falls into group III, test it for streak and note other general and specific properties such as color, hardness, cleavage, and so on, until the mineral fits the description of one of those given in table 1.2 under group III.

4. To assist you in confirming your identification, refer to the expanded mineral descriptions in table 1.3.

5. Your laboratory instructor will advise you as to the procedure to be used to verify your identification.

6. Refer to table 1.3 to learn about occurrence and economic value of each mineral. The chemical groupings and composition of some of the common minerals are presented in table 1.4.

Odor

Some minerals give off a characteristic odor when damp. Exhaling on the kaolinite specimen, thus dampening it, causes that mineral to exude a musty or dank odor.

Feel

The feel of a mineral is the impression gained by handling or rubbing it. Terms used to describe feel are common descriptive adjectives such as soapy, greasy, smooth, rough, and so forth.

Chemical Reaction

Calcite will effervesce (bubble) when treated with cold dilute (0.1N) hydro-chloric acid. NOTE: Your laboratory instructor will provide the proper dilute acid if you are to use this test.

Table 1.2   Mineral Identification Key

I. Non-metallic, light-colored	**Hard (scratches glass)**	**Shows cleavage**	Vitreous luster. Color varies from white or cream to pink. Hardness 6. Cleavage two planes at nearly 90 degrees. Streak white. G=2.56. Crystals uncommon. Grains have glossy appearance.	ORTHOCLASE
			Vitreous luster. Color varies from white to gray or reddish to reddish brown. Hardness 6. Cleavage two planes at nearly 90 degrees. Cleavage planes show striations. Streak white. G=2.6–2.75. Striations diagnostic. Some samples may show a play of colors.	PLAGIOCLASE
		No cleavage	Vitreous luster. Colorless or white, but almost any color can occur. Hardness 7. Cleavage none. Conchoidal fracture. Streak white. G=2.65. Hexagonal crystals with striations common. Also massive.	QUARTZ
			Waxy or dull luster. Color varies from white to pale yellow, brown, or gray. Hardness 7. Cleavage none. Streak white. G=2.65. Characterized by conchoidal fracture with sharp edges.	CHALCEDONY (flint/chert)
			Waxy luster. Variegated banded colors. Hardness 7. Cleavage none. Streak white.	CHALCEDONY (agate)
			Vitreous luster. Color commonly olive-green, sometimes yellowish. Hardness 6.5–7. Cleavage indistinct. Streak white or gray. G=6.5–7. Commonly in granular masses.	OLIVINE
	Soft (does not scratch glass)	**Shows cleavage**	Vitreous luster. Colorless, also white, gray, yellow, or red. Hardness 2.5. Perfect cubic cleavage. Streak white. G=2.5. Table salt taste.	HALITE
			Vitreous luster. Colorless and transparent, white, variety of color possible. Hardness 3. Perfect rhombohedral cleavage. Streak white to gray. G=2.7. Effervesces in cold dilute HCl; double refraction in colorless varieties.	CALCITE
			Vitreous to pearly luster. Colorless, white, gray, greenish, or yellow-brown. Hardness 3.5–4. Rhombohedral cleavage. Streak white. G=2.85–3.2. Crystals common, twinning common. Reaction with cold dilute HCl only when powdered.	DOLOMITE
			Vitreous to pearly luster. Colorless to white, gray, yellowish orange, or light brown. Hardness 2. Cleavage good in one direction producing thin sheets. Fracture may be fibrous. Streak white. G=2.32. Crystals common, twinning common.	GYPSUM
			Pearly to greasy to dull luster. Color usually pale green, also shades of white or gray. Hardness 1. Perfect basal cleavage. Streak white. G=2.82. Soapy feel.	TALC
			Vitreous to silky or pearly luster. Colorless to shades of green, gray, or brown. Hardness 2.5–4. Perfect basal cleavage yielding thin flexible and elastic sheets. Streak white. G=2.8–2.9.	MUSCOVITE
			Greasy waxlike to silky luster. Color variable, shades of green most common. Hardness 2–5. Cleavage none. Fibrous parting. Streak white. G=2.5–2.6.	ASBESTOS
			Vitreous luster. Colorless but wide range of colors possible. Hardness 4. Perfect octahedral cleavage (4 planes). Streak white. G=3.18.	FLUORITE
		No cleavage	Dull to earthy luster. Color white, often stained. Hardness 2. No cleavage apparent in common massive varieties. Streak white. G=2.6. Earthy smell when damp.	KAOLINITE
			Pearl to greasy or dull luster. Color pale green or shades of gray. Hardness 1. No apparent cleavage in massive varieties. Streak white. G=2.82. Soapy feel.	TALC
			Earthy luster. White or various colors. Hardness 3, may be less. No apparent cleavage. Streak white, G=2.7. Effervesces in cold dilute HCl.	CALCITE
			Earthy luster. White or various colors. Hardness 3.5–4, but apparent may be less. No apparent cleavage. Streak white. Reacts with cold dilute HCl only when powdered.	DOLOMITE
			Earthy luster. White color. Hardness 2, but apparent may be less. No apparent cleavage. Streak white. G=2.32. Massive fine-grained variety called *alabaster*.	GYPSUM

Table 1.2   Continued

			Description	Mineral
II. Non-metallic, dark-colored	**Hard (scratches glass)**	**Shows cleavage**	Vitreous luster. Color black. Hardness 5–6. Cleavage two planes at nearly 90 degrees. Streak white to gray. G=3.2–3.6. May exhibit parting.	AUGITE
			Vitreous luster. Color black. Hardness 5–6. Cleavage two planes with intersections at 56 and 124 degrees. G=3–3.4. Six-sided crystals common.	HORN-BLENDE
			Vitreous luster. Color ranges from white, gray, to reddish or reddish brown. Hardness 6. Cleavage two planes at nearly 90 degrees. Striations on cleavage planes. Streak white. G=2.6–2.75. Some forms exhibit play of colors on cleavage surfaces.	PLAGIO-CLASE
		No cleavage	Adamantine to vitreous luster. Color varies but commonly brown. Hardness 9. Cleavage none. G=4.0. Barrel-shaped hexagonal crystals with striations on basal faces common.	CORUNDUM
			Vitreous to resinous luster. Color varies but dark red to reddish brown common. Hardness 6.5–7.5. Cleavage none. Streak white or shade of the mineral color. G=3.6–4.3. Fracture may resemble a poor cleavage. Brittle.	GARNET
			Vitreous luster. Color commonly olive-green, sometimes yellowish. Hardness 6.5. Cleavage indistinct. Streak white or gray. G=3.2–3.4. Commonly in granular masses.	OLIVINE
			Vitreous luster. Color gray to gray-black. Hardness 7. Cleavage none. Streak white. G=2.65. Conchoidal fracture. Crystal common; also a variety of massive forms.	QUARTZ
			Waxy to dull luster. Color red to red-brown or brown. Hardness 7. Cleavage none. Streak white to gray. G=2.6.	CHALCEDONY (jasper)
			Waxy or dull luster. Color dark gray to black. Hardness 7. Cleavage none. Streak white. G=2.6. Characterized by conchoidal fracture with sharp edges.	CHALCEDONY (flint/chert)
	Soft (does not scratch glass)	**Shows cleavage**	Vitreous to pearly luster. Color dark green, brown, to black. Hardness 2.5–4.0. Perfect basal cleavage forming thin elastic sheets. Streak white to gray. G=2.9–3.1.	BIOTITE
			Resinous luster. Color yellow brown to dark brown. Hardness 3.5–4.0. Cleavage perfect in six directions (dodecahedral). Sreak brown to light yellow or white. G=3.9–4.1. Cleavage faces common; twinning common.	SPHALERITE
			Vitreous to earthy luster. Color green to greenish black. Hardness 2.5. Perfect basal cleavage forming flexible nonelastic sheets. Streak white to pale green. G=2.7–3.3. May have slippery feel.	CHLORITE
		No cleavage	Submetallic to earthy luster. Color red to red-brown. Hardness 5–6 but apparent may be as low as 1. Cleavage none. Streak red. G=5.0–6.0. Earthy appearance.	HEMATITE (soft iron ore)
			Vitreous to subresinous luster. Color variable; green, blue, brown, purple. Hardness 5. Cleavage poor basal. Streak white. G=3.15–3.2. Crystals common.	APATITE
			Earthy luster. Color variable; yellow, yellow-brown to brownish black. Apparent hardness 1. No cleavage apparent in earthy masses. Streak brownish yellow to orange-yellow. G=3.3–3.4. Earthy masses.	GOETHITE (limonite)
III. Metallic luster	**Black, green-black, or dark green streak**		Metallic luster. Color black. Hardness 6. Cleavage none. Streak black. G=5.2. Strongly magnetic.	MAGNETITE
			Metallic luster. Color dark gray to black. Hardness 1–2. Perfect basal cleavage. Streak black. G=2.1–2.25. Greasy feel, smudges fingers when handled.	GRAPHITE
			Metallic luster. Hardness 6–6.5. Cleavage none. Streak greenish or brownish black. Cubic crystals with striated faces common.	PYRITE
			Metallic luster. Color brass-yellow, often tarnished to bronze or purple. Hardness 3.5–4. Cleavage none. Streak greenish black. G=4.1–4.3. Usually massive.	CHALCO-PYRITE
			Bright metallic luster. Color shiny lead-gray. Hardness 2.5. Perfect cubic cleavage. Streak lead-gray. G=7.5–7.6. Cleavage, high G, and softness diagnostic.	GALENA
	Red streak		Metallic luster. Color steel-gray. Hardness 5–6. Cleavage none. Streak red to red-brown. G=5.6. Often micaceous or foliated. Brittle.	HEMATITE (specularite)
	Yellow, brown, or white streak		Metallic to dull luster. Color yellow-brown to dark brown, may be almost black. Hardness 6. Cleavage perfect parallel to side pinacoid. Streak brownish yellow to orange-yellow. G=3.3–4.3. Brittle.	GOETHITE
			Submetallic to resinous luster. Color yellow to yellow-brown to dark brown. Hardness 3.5–4. Perfect cleavage in six directions (dodecahedral). Streak brown to light yellow to white. G=3.9–4.1. Cleavage faces common; twinning common.	SPHALERITE

Figure 1.5 Quartz crystal.

Figure 1.6 Rose quartz.

Figure 1.7 Smoky quartz.

Figure 1.8 Cryptocrystalline quartz (chert).

Figure 1.9 Orthoclase (microcline).

Figure 1.10 Plagioclase.

Figure 1.11 Gypsum (selenite).

Figure 1.12 Talc.

Figure 1.13   Calcite (note double refraction).

Figure 1.14   Fluorite.

Figure 1.15   Biotite.

Figure 1.16   Olivine.

Figure 1.17   Hematite (specularite).

Figure 1.18   Hematite.

Figure 1.19   Goethite (limonite).

Figure 1.20   Pyrite.

Table 1.3   Mineral Catalog

APATITE Ca, F Phosphate Hexagonal	Luster nonmetallic; vitreous to subresinous. Color variable; greenish yellow, blue, green, brown, purple, white. Hardness 5. Cleavage poor basal; fracture conchoidal. Streak white. G=3.15–3.2 . Crystals common; also, massive or granular forms. Important source of phosphorus for fertilizers, phosphoric acid, detergents, and munitions.
ASBESTOS (Serpentine) Mg, Al Silicate Monoclinic	Luster nonmetallic; greasy or waxlike in massive varieties, silky when fibrous. Color variable; shades of green most common. Hardness 2–5. Cleavage none. Streak white. G=2.5–2.6. Occurs in massive, platy, and fibrous forms. Wide industrial uses, especially in roofing materials.
AUGITE Ca, Mg, Fe, Al Silicate Monoclinic	Luster nonmetallic; vitreous. Color dark green to black. Hardness 6. Cleavage good, two planes at nearly 90°; may exhibit well-developed parting. Streak white to gray. G=3.2–3.6. Short stubby eight-sided prismatic crystals. Often in granular crystalline masses. Most important ferromagnesium mineral in dark igneous rocks. No commercial value. *Note:* The *pyroxene* group of minerals contains over 15 members. Augite is an example of the monoclinic members of this group.
BIOTITE K, Mg, Fe, Al Silicate Monoclinic	Luster nonmetallic; vitreous to pearly. Color dark green, brown, or black. Hardness 2.5–4. Cleavage perfect basal forming elastic sheets. Streak white to gray. G=2.9–3.4. Crystals common as pseudohexagonal prisms, more commonly in sheets or granular crystalline masses. Common accessory mineral in igneous rocks, also important in some metamorphic rocks. Commercial value as insulator and in electrical devices.
CALCITE $CaCO_3$ Trigonal	Luster nonmetallic; vitreous (may appear earthy in fine-grained massive forms). Colorless and transparent or white when pure; wide range of colors possible. Hardness 3. Cleavage perfect rhombohedral. Streak white to gray. G=2.7. Crystals common. Also massive, granular, oolitic, or in a variety of other habits. Effervesces in cold dilute HCl. Strong double refraction in colorless varieties. Common and widely distributed rock-forming mineral in sedimentary and metamorphic rocks. Chief raw material for portland cement; wide variety of other uses.
CHALCEDONY (Cryptocrystalline Quartz) SiO_2 Trigonal	Luster nonmetallic; waxy or dull. (Chalcedony is a group name for a variety of forms of extremely fine grained, very diverse quartz including *agate*, banded forms; *carnelian* or *sard*, red to brown; *jasper*, opaque and generally red or brown; *chert* and *flint*, massive, opaque and ranging in color from white, pale yellow, brown, gray, to black and characterized by conchoidal fracture with sharp edges; *silicifed wood*, reddish or brown showing wood structure.) Hardness 7. Cleavage none. Streak white. G=2.6. Occurs in a wide variety of sedimentary rocks and veins, cavities, or as dripstone. Various forms used as semiprecious stones.
CHALCOPYRITE $CuFeS_2$ Trigonal	Luster metallic, opaque. Color brass-yellow, often tarnished to bronze or purple. Hardness 3.5–4. Cleavage none, fracture uneven. Streak greenish black. G=4.1–4.3. May occur as small crystals but usually massive. Most important and common copper ore mineral.
CHERT (see Chalcedony)	
CHLORITE Mg, Fe, Al Silicate Monoclinic	Luster nonmetallic; vitreous to earthy. Color green to green-blue. Hardness 2.5. Cleavage perfect basal forming flexible nonelastic sheets. Streak white to pale green. G=2.7–3.3. Occurs as foliated masses or small flakes. Common in low-grade schists and as alteration product of other ferromagnesian minerals. No commercial value.

Table 1.3 Continued

CORUNDUM Al$_2$O$_3$ Trigonal	Luster nonmetallic, adamantine to vitreous. Color varies, yellow, brown, green, purple; gem varieties blue (sapphire) and red (ruby). Hardness 9. Cleavage none, parting common with striations on parting planes. Streak white, G=4. Barrel-shaped crystals common, frequently with deep horizontal striations. Wide uses as an abrasive.
DOLOMITE CaMg(CO$_3$)$_2$ Trigonal	Luster nonmetallic, vitreous to pearly. Colorless, white, gray, greenish, yellow-brown; other colors possible. Hardness 3.5–4. Cleavage rhombohedral. Streak white. G=2.85–3.2. Crystals common. Twinning common. Fine grained; massive and granular forms common also. Distinguished from calcite by fact that it effervesces in cold dilute HC1 only when powdered. Widespread occurrence in sedimentary rocks. Variety of industrial uses as flux, as source of magnesia for refractory bricks, and as a source of magnesium or calcium metal.
FLINT/JASPER (see Chalcedony)	
FLUORITE CaF$_2$ Cubic (Isometric)	Luster nonmetallic, vitreous transparent to translucent. Colorless when pure; occurs in a wide variety of colors: yellow, green, blue, purple, brown, and shades in between. Hardness 4. Cleavage perfect octahedral (four directions parallel to faces of an octahedron). Streak white. G=3.18. Twins fairly common. Common as a vein mineral. Industrial use as a flux in steel and aluminum metal smelting; source of fluorine for hydrofluoric acid.
GALENA PbS Cubic (Isometric)	Luster bright metallic. Color lead-gray. Hardness 2.5. Cleavage perfect cubic. Streak lead-gray. G=7.5–7.6. Easily identified by cleavage, high specific gravity, and softness. Chief lead ore.
GARNET Fe, Mg, Ca, Al Silicate Cubic (Isometric)	Luster nonmetallic; vitreous to resinous. Color varies but dark red and reddish brown most common; white, pink, yellow, green, black depending on composition. Hardness 6.5–7.5. Cleavage none. Streak white or shade of mineral color. G=3.6–4.3. Crystals common but also in granular masses. Some value as an abrasive. Gemstone varieties are pyrope (red) and andradite (green).
GOETHITE FeO (OH) Orthorhombic	Luster variable, crystals adamantine; metallic to dull in masses, fibrous varieties may be silky. Color dark brown, yellow-brown, reddish brown, brownish black, yellow. Hardness 5–5.5. Cleavage perfect parallel to side pinacoid, fracture uneven. Streak brownish yellow to orange-yellow. G=3.3–4.3. Crystals uncommon. Usually massive or earthy as residual from chemical weathering, or stalactic by direct precipitation. Often in radiating fibrous forms. This species includes the common brown and yellow-brown ferric oxides collectively called *limonite*. Commercial source of iron ore.
GRAPHITE C Hexagonal	Luster metallic to dull. Color dark gray to black. Hardness 1–2. Cleavage perfect basal. Streak black. G=2.1–2.25. Characteristic greasy feel, marks easily on paper. Crystals uncommon, usually as foliated masses. Common metamorphic mineral. Wide industrial uses due to high melting temperature (3000°C) and insolubility in acid. Used also as lead in pencils.
GYPSUM CaSO$_4$·2H$_2$O Monoclinic	Luster nonmetallic; vitreous to pearly; some varieties silky. Colorless to white, gray, yellowish orange, light brown. Hardness 2. Cleavage good in one direction producing thin sheets; fracture conchoidal in one direction, fibrous in another. Streak white. G=2.32. Crystals common and simple in habit; twinning common. Varieties include *selenite*, coarsely crystalline, colorless and transparent; *satinspar*, parallel fibrous structure; and *alabaster*, massive fine-grained gypsum. Occurs widely as sedimentary deposits and in many other ways. Mined for use in wallboard, plaster, and filler for paper products.

Table 1.3   Mineral Catalog, continued

HALITE NaCl Cubic (Isometric)	Luster nonmetallic, vitreous. Transparent to translucent. Colorless, also white, gray, yellow, red. Hardness 2.5. Cleavage perfect cubic. Streak white. G=2.5. Crystals common, also massive or coarsely granular. Characteristic taste of table salt. Widely used as source of both sodium and chlorine and as salt for table, pottery glaze, and industrial purposes.
HEMATITE Fe_2O_3 Trigonal	Luster metallic in form known as specularite and in crystals; submetallic to dull in other varieties. Color steel-gray in specularite, dull to bright red in other varieties. Hardness 5–6, but apparent may be as low as 1. Cleavage none; basal parting fracture uneven. Streak red-brown. G=5–6. Crystals uncommon. May occur in crystalline, botryoidal, or earthy masses. Specularite commonly micaceous or foliated. Streak characteristic. Common iron ore.
HORNBLENDE (Amphibole) Ca, Na, Mg, Fe, Al Silicate Monoclinic	Luster nonmetallic, vitreous. Color dark green, dark brown, black. Hardness 6. Cleavage perfect on two planes meeting at 56° and 124°. Streak gray or pale green. G=3–3.4. Long, six-sided crystals common. Color usually darker than other minerals in amphibole group. No commercial use. *Tremolite-Actinolite*, also a member of the amphibole group, is lighter in color and commonly is fibrous or asbestiform, may range in color from white to green, and has white streak. *Nephrite*, the amphibole form of jade, is used for jewelry.
KAOLINITE Hydrous Aluminosilicate Triclinic	Luster nonmetallic, dull to earthy. Color white, often stained by impurities to red, brown, or gray. Hardness 2. Cleavage perfect basal but rarely seen because of small grain size. Streak white. G=2.6. Found in earthy masses. Earthy odor when damp. *NOTE*: Kaolinite is used here as an example of the clay minerals. It normally is not possible to distinguish the various clay minerals on the basis of their physical properties. Other clay minerals include *montmorillonite* (smectite), *illite*, and *vermiculite*. Kaolinite is used as a paper filler and ceramic, montmorillonite for drilling muds, illite has no industrial use, and vermiculite is mined and processed for use as a lightweight aggregate, potting soils, and as insulation.
LIMONITE (see Goethite)	
MAGNETITE $FeFe_2O_4$ Cubic (Isometric)	Luster metallic. Color black. Hardness 6. Cleavage none, some octahedral parting. Streak black. G=5.2. Usually in granular masses. Strongly magnetic, some specimens show polarity (lodestones). Widespread occurrence in a variety of rocks. Used commercially as iron ore.
MUSCOVITE K, Al Silicate Monoclinic	Luster nonmetallic, vitreous to silky or pearly. Colorless to shades of green, gray, or brown. Hardness 2.5–4. Cleavage perfect basal yielding thin sheets that are flexible and elastic; may show some parting. Streak white. G=2.8–2.9. Usually in small flakes or lamellar masses. Commercial deposits found in granite pegmatites but occur in many other rocks. Variety of industrial uses.
OLIVINE $(Mg, Fe)_2SiO_4$ Orthorhombic	Luster nonmetallic, vitreous. Color olive-green to yellowish; nearly pure Mg-rich varieties may be white (forsterite) and nearly pure Fe-rich varieties brown to black (fayalite). Hardness 6.5. Cleavage indistinct. Streak white or gray. G=3.2–3.4. Usually in granular masses. Crystals uncommon. A mineral of basic and ultrabasic igneous rocks. Forsterite variety used for refractory bricks.

Table 1.3   Continued

ORTHOCLASE (K-Feldspar) $K(AlSi_3O_8)$ Monoclinic	Luster nonmetallic, vitreous. Color varies, white, cream, or pink; *sanidine* variety may be colorless. Hardness 6. Cleavage two planes at nearly right angles. Streak white. G=2.56. Crystals not common. Has glossy appearance. Distinguished from other feldspars by absence of twinning striations. *NOTE: Microcline* variety is triclinic. When light green the color is diagnostic, more commonly white, green, pink. Occurrence helpful; most K-feldspar in pegmatites is microcline.
PLAGIOCLASE Ranges in composition from Albite, $NaAlSi_3O_8$, to Anorthite, $CaAl_2Si_2O_8$ Triclinic	Luster nonmetallic, vitreous. Color white or gray, reddish, or reddish brown. Hardness 6. Cleavage two planes at close to right angles, twinning striations common on basal cleavage surfaces. Streak white. G=2.6–2.75. Crystals common for Na-rich varieties, uncommon for intermediate varieties, rare for anorthite. Twinning common. Twinning striations on basal cleavage useful to distinguish from orthoclase. Some varieties show play of colors. Sodium-rich varieties mined for use in ceramics.
PYRITE FeS_2 Cubic (Isometric)	Luster metallic. Color brass-yellow, may be iridescent if tarnished. Hardness 6–6.5. Cleavage none, conchoidal fracture. Streak greenish or brownish black. G=5.0. Crystals common, usually cubic with striated faces. Crystals may be deformed. Massive granular forms also. Most widespread sulfide mineral. Known as "fool's gold." Source of sulfur for sulfuric acid. *NOTE: Marcasite* (FeS_2) is orthorhombic, usually paler in color, and is commonly altered.
QUARTZ SiO_2 Trigonal	Luster nonmetallic, vitreous. Typically colorless or white, but almost any color may occur. Hardness 7. Cleavage none, conchoidal fracture. Streak white but difficult to obtain on streak plate. G−2.65. Prismatic crystals common with striations perpendicular to the long dimension; also a variety of massive forms. Color variations lead to varieties called smoky quartz, rose quartz, milky quartz, and amethyst. Common mineral in all classes of rocks. Wide variety of commercial uses including glassmaking, electronics, as a flux, and construction products.
SPHALERITE ZnS Cubic (Isometric)	Luster usually nonmetallic, some varieties submetallic, most commonly resinous. Color yellow, yellow-brown to dark brown. Hardness 3.5–4. Cleavage perfect dodecahedral (six directions at 120°) and common. Streak brown to light yellow or white. G=3.9–4.1. Crystals common as distorted tetrahedra or dodecahedra. Twinning common. Also massive or granular. Important zinc ore.
TALC Mg Silicate Monoclinic	Luster nonmetallic, pearly to greasy or dull. Usually pale green, also white to silver-white or gray. Hardness 1. Cleavage perfect basal, massive forms show no visible cleavage. Streak white. G=2.82. Usually foliated masses or dense fine-grained dark gray to green aggregates (soapstone). Crystals extremely rare. Soapy feel is diagnostic. Commercial uses in paints, ceramics, roofing, paper, and toilet articles.

Table 1.4. Chemical Grouping and Composition of Some Common Minerals

EXAMPLE		
CHEMICAL GROUP	**MINERAL NAME**	**CHEMICAL FORMULA*†**
ELEMENTS	Native copper	Cu
	Graphite	C
	Diamond	C
OXIDES	Quartz	SiO_2
	Hematite	Fe_2O_3
	Magnetite	$FeFe_2O_4$
	Goethite	$FeO(OH)$
	Corundum	Al_2O_3
SULFIDES	Pyrite	FeS_2
	Chalcopyrite	$CuFeS_2$
	Galena	PbS
	Sphalerite	ZnS
SULFATES	Anhydrite	$CaSO_4$
	Gypsum	$CaSO_4 \cdot 2H_2O$
CARBONATES	Calcite	$CaCO_3$
	Dolomite	$CaMg(CO_3)_2$
PHOSPHATES	Apatite	$Ca_5(PO_4)_3F$
HALIDES	Halite	$NaCl$
	Fluorite	CaF_2
SILICATES — OLIVINE GROUP	Olivine	$(Mg,Fe)_2SiO_4$
SILICATES — AMPHIBOLE GROUP	Hornblende	Ca,Na,Mg,Fe,Al Silicate
	Asbestos (fibrous serpentine)	Mg,Al Silicate
SILICATES — PYROXENE GROUP	Augite	Ca,Mg,Fe,Al Silicate
SILICATES — MICA GROUP	Muscovite Biotite Chlorite Talc Kaolinite	K,Al Silicate K,Mg,Fe,Al Silicate Mg,Fe,Al Silicate Mg Silicate Al Silicate
SILICATES — FELDSPAR GROUP	Orthoclase (K-feldspar) Plagioclase (Ab,An) Albite (Ab) Anorthite (An)	$K(AlSi_3O_8)$ Mixture of Ab and An $NaAlSi_3O_8$ $CaAl_2Si_2O_8$

*Some common elements and their symbols:

Al—Aluminum	Fe—Iron	O—Oxygen
C—Carbon	H—Hydrogen	P—Phosphorus
Ca—Calcium	K—Potassium	Pb—Lead
Cl—Chlorine	Mg—Magnesium	S—Sulfur
Cu—Copper	Mn—Manganese	Si—Silicon
F—Fluorine	Na—Sodium	Zn—Zinc

†Chemical formulas from *Mineralogy*, 2d ed., by L. G. Berry, Brian Mason, and R. V. Dietrich (San Francisco: W. H. Freeman, 1983).

Rocks

Background

As a part of the study of rocks in hand specimen or in the field, some understanding of the origin and occurrence of the rocks being studied is essential. The first part of this section deals with the identification and classification of rocks based on the study of hand specimens with reference also to their origin and occurrence. The second part of the section introduces some basic principles dealing with the relationships between rock types in their natural setting. These principles will be used in understanding the relative ages of adjacent rock masses; thus the relationships between rocks can help in tracing the geologic history of a given area. Several simple diagrams and schematic drawings will be used to convey the ideas and concepts.

Three major categories of rocks are recognized. They are *igneous,* rocks formed by the cooling and crystallization of molten material within or at the surface of the earth; *sedimentary,* rocks formed from sediments derived from preexisting rocks, by precipitation from solution, or by the accumulation of organic materials; and *metamorphic,* rocks resulting from the change of preexisting rocks into new rocks with different textures and mineralogy as a result of the effects of heat, pressure, chemical action, or combinations of these. All types of these three categories of rocks can be observed at the earth's surface today. We can also observe some of the processes that lead to their formation. For example, the eruption of volcanoes produces certain types of igneous rocks. We can observe the weathering, erosion, transportation, and deposition of sediments that upon *lithification*—the combination of processes that convert a sediment into an indurated rock—produce certain types of sedimentary rocks. Other observable processes of rock formation at the surface of the earth include the deposition of dripstone in caves and the growth of coral reefs in the oceans.

The earth's crust is dynamic and is subject to a variety of processes that act upon these three rock types. Given time and the effect of these processes, any one of these rocks can be changed into another type. This relationship is the basis of the *rock cycle,* represented in figure 1.21. The heavy arrows indicate the normal complete cycle,

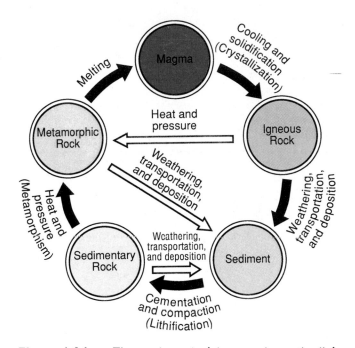

Figure 1.21   The rock cycle (shown schematically).

while the open arrows indicate how this cycle may be interrupted. Keep in mind that the schematic presented in figure 1.21 is greatly simplified and that a given rock observed today represents only the last phase of the cycle from which it has been derived.

Igneous Rocks

Origin

Igneous rocks are aggregates of minerals that crystallize from a molten material that is generated deep within the earth's mantle. The heat required to generate this melted material comes from within the earth. The temperature of the earth increases with depth at a rate of about 30° C per kilometer of depth. This increase in temperature is known as the *geothermal gradient,* and while it varies from

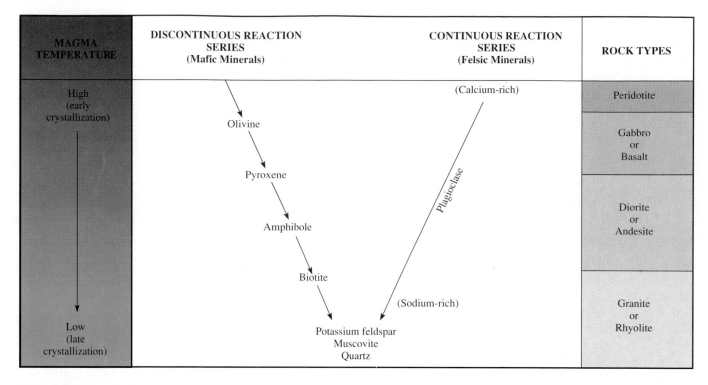

Figure 1.22   Reaction series for igneous rock formation from a magma.

place to place, at a depth of about 35 kilometers the temperature is sufficient to melt rock. The molten material, *magma,* is a complex solution of silicates plus water and various gases. Some of this magma may reach the surface of the earth where it is extruded as *lava,* but other magmas solidify before they reach the surface. The rocks formed by solidification of magma within the mantle or crust are called *intrusive* igneous rocks, and those that form at the surface from lavas are called *extrusive* igneous rocks.

The composition of a given magma depends on the composition of the rocks that were melted to form it. Once melt occurs, the magma tends to rise toward the surface of the earth. As it rises, cooling and crystallization begin. Ultimately, the molten mass solidifies into solid rock. The type of igneous rock formed depends on a number of factors including the original composition of the melt, the rate of cooling, and the reactions that occurred within the magma as cooling took place.

Intrusive rock masses have been studied in detail, and it is recognized that there is an orderly sequence in which the various minerals crystallize. This sequence of crystallization (commonly called *Bowen's Reaction Series* after the geologist who first proposed it) has been verified in principle.

The reaction series presented in figure 1.22 gives some suggestion as to the formation of the various types of igneous rocks, and it helps to explain why the association of some minerals (such as olivine and quartz) is rare in nature. Study of figure 1.22 indicates that the first minerals to form tend to be low in silica. The plagioclase side

of the diagram is labeled as a *continuous reaction series.* This means that the first crystals to form, calcium-rich plagioclase (anorthite), continue to react with the remaining melt as cooling continues. This process involves the substitution of sodium for calcium in the plagioclase without a change in the crystal structure. As this process continues, the composition of the plagioclase becomes increasingly sodium- and silica-rich, and the last plagioclase formed is albite.

The *discontinuous series* consists of the common ferromagnesian minerals found in igneous rocks. In this series the first mineral to crystallize is olivine. As the cooling continues, the olivine crystals react with the remaining melt and form pyroxenes at the expense of olivine. As this process of interaction between the crystals and the silica-rich melt continues, the pyroxene reacts to form amphibole, and the final member of this discontinuous series is biotite, which forms at the expense of amphibole.

If in fact the original magma is low in silica and high in iron and magnesium, the magma may solidify before the complete series of reactions has occurred. The resulting rocks are high in magnesium and iron, low in silica, and are said to be *mafic.* Olivine, pyroxene, and calcium-rich plagioclase are the common mineral associations of mafic rocks. Conversely, melts originally high in silica and low in ferromagnesian elements may reach the final stages of the reaction series, and the rocks formed are composed of potassium feldspar, quartz, and muscovite. These rocks are said to be *felsic* or *sialic.*

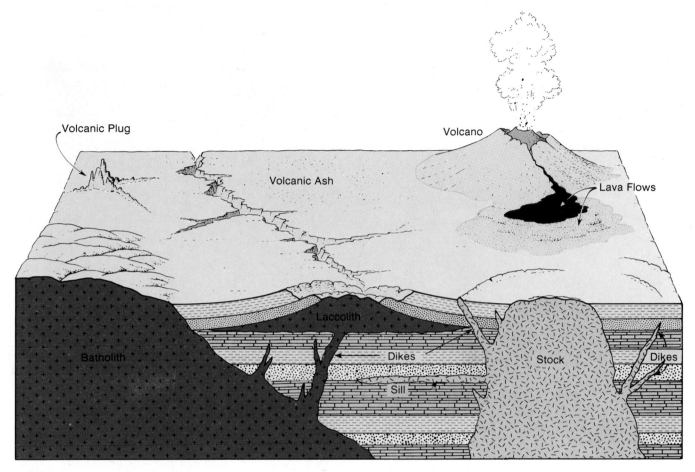

Figure 1.23 Block diagram showing various modes of occurrence of igneous rocks.

Keep in mind that this reaction series is idealized. Changes in the original composition of the melt occur in nature by *fractional crystallization,* the removal of crystals from the melt by settling or filtering, by *assimilation* of part of the rock through which it is rising, or by the *mixing* of two magmas of differing composition.

Occurrence

As magma works its way toward the surface of the earth it encounters solid preexisting rock. The rock mass intruded by the magma is referred to as *country rock,* and it is not uncommon for pieces of the country rock to be engulfed by the magma. Fragments of country rock surrounded by igneous rocks are called *inclusions.*

Intrusive and extrusive igneous rocks may assume any number of geometric forms. Figure 1.23 illustrates several of the more common shapes of igneous rock masses found in nature, and the relationship of igneous rock types to their mode of occurrence is shown in table 1.5.

The largest intrusive igneous rock mass (*pluton*) is a *batholith.* Batholiths are usually coarse grained (*phaneritic*) in texture and granitic in composition. By definition,

Table 1.5. Relationship of Igneous Rock Types to Their Modes of Occurrence in the Earth's Crust.

	ROCK TYPE	SOME MODES OF OCCURRENCE
EXTRUSIVE	Pumice Scoria	Lava flows, pyroclastics Crusts on lava flows, pyroclastics
	Obsidian	Lava flows
	Rhyolite Andesite Basalt	Lava flows, shallow intrusives
INTRUSIVE	Rhyolite porphyry Andesite porphyry Basalt porphyry	Dikes, sills, laccoliths, intruded at medium to shallow depths
	Granite Diorite Gabbro Peridotite	Batholiths and stocks of deep-seated intrusive origin

Figure 1.24   Basaltic dike intruding Precambrian granite north of Lake Superior on the Canadian Shield.

a batholith crops out over an area of more than 100 square kilometers (about 36 square miles). A *stock* is similar in composition but, by definition, crops out over an area less than 100 kilometers (less than 36 square miles). A *dike* is a tabular igneous intrusion whose contacts cut across the trend of the country rock (fig. 1.24). A *sill* is also tabular in shape, but its contacts lie parallel to the trend of the country rock. A *laccolith* is similar to a sill but is generally much thicker, especially near its center where it has caused the country rock to bulge upward. These five major types of plutons—batholiths, stocks, dikes, sills, and laccoliths—can be grouped into two main categories based on the relationship of their contacts to the trend of the enclosing country rock. The contacts of batholiths, stocks, and dikes all cut across the trend of the country rock and hence are called *discordant igneous plutons*. Sills and laccoliths, on the other hand, have contacts that are parallel to the trends of the country rock and are called *concordant igneous plutons*. Table 1.6 summarizes these relationships.

Table 1.6.   Relationship between Concordant or Discordant Igneous Bodies and Their Modes of Occurrence

RELATIONSHIP OF IGNEOUS ROCK CONTACT TO COUNTRY ROCK	MODE OF OCCURRENCE
CONCORDANT	SILL, LACCOLITH
DISCORDANT	BATHOLITH, STOCK, DIKE

Textures of Igneous Rocks

The *texture* of a rock is its appearance that results from the size, shape, and arrangement of the mineral grains or crystals in the rock. The texture of igneous rocks can be described in terms of one of the following:

Phaneritic Texture or Coarse Grained

This term applies to an igneous rock in which the constituent minerals are macroscopic in size. The dimensions of the individual crystals or grains range from about 1 mm to more than 5 mm (figs. 1.25–1.29).

Aphanitic Texture or Fine Grained

This term is used to describe the texture of an igneous rock composed of mineral crystals or grains that are microscopic in size; that is, they cannot be discerned with the naked or corrected eye (fig. 1.30).

Porphyritic Texture

This texture is characteristic of an igneous rock in which the macroscopic mineral crystals or grains are embedded in a matrix of microscopic crystals or grains (figs. 1.32 and 1.34), or macroscopic crystals or grains of one size range occur in a matrix of smaller macroscopic crystals or grains. The larger crystals or grains are called *phenocrysts,* and the smaller grains in which they are embedded are called the *groundmass*.

Vesicular Texture

This texture is characterized by the presence of *vesicles*—tubular, ovoid, or spherical cavities in the rock (fig. 1.31). These "holes" in the rock are a result of gas bubbles being trapped in the rock as it cools. The size of the vesicles range from less than 1 mm up to several centimeters in diameter. Vesicles that are filled with mineral matter are called *amygdules,* and the texture then is called *amygdaloidal* texture.

Glassy Texture

This texture resembles that of glass (fig. 1.39).

The texture of igneous rocks is a reflection of the mineralogy and the cooling history of the magma or lava from which they were formed. Igneous rocks with a phaneritic or coarse-grained texture were formed where the cooling rate was very slow and larger crystals or grains could form. Igneous rocks with an aphanitic or fine-grained texture were formed where cooling was much more rapid. A porphyritic texture reflects a two-stage cooling history. The larger crystals formed first under conditions of slow cooling, but before the magma or lava turned to rock it migrated to a zone of faster cooling where the remainder of the melt solidified. Vesicular textures are indicative of lava flows in which escaping gases produced the vesicles while the lava was still molten. A glassy texture reflects an extremely rapid rate of cooling in the absence of gases. Lava flows commonly possess a glassy texture.

Mineralogic Composition of Igneous Rocks

The minerals of igneous rocks are grouped into two major categories, *primary* and *secondary*. Primary minerals are those that are crystallized from the cooling magma. Secondary minerals are those formed after the magma had solidified, and include minerals formed by chemical alteration of the primary minerals, or by the deposition of new minerals in an igneous rock. *Secondary minerals are not important in the classification of igneous rocks.*

Primary minerals consist of two types for the purpose of classifying igneous rocks—*essential minerals* and *accessory minerals*. The essential minerals are those that must be present in order for the rock to be assigned a specific position in the classification scheme (i.e., given a specific name). Accessory minerals are those that may or may not be present in a rock of a given type, but the presence of an accessory mineral in a given rock may affect the name of the rock. For example, the essential minerals of a granite are quartz and K-feldspar (K-feldspar is a common notation for the group of potassium-rich feldspars of which orthoclase is the most common). If a particular granite contains an accessory mineral such as biotite or hornblende, the rock may be called a biotite granite or a hornblende granite, respectively.

The essential minerals contained in the common igneous rocks are quartz, K-feldspar, plagioclase, pyroxene (commonly augite), amphibole (commonly hornblende), and olivine. The key to igneous rock identification is the ability of the observer to recognize the presence or absence of quartz and to distinguish between K-feldspar and plagioclase. Color is of little help in the matter because quartz, K-feldspar, and plagioclase can all occur in the same shade of gray. The distinction between quartz and the feldspars is made by the fact that quartz has no cleavage; macroscopic quartz crystals do not exhibit shiny cleavage faces as the feldspars do.

The distinction between K-feldspar and plagioclase is more difficult because both have cleavage faces that show up as shiny surfaces in phaneritic hand specimens. A pink-colored feldspar is usually K-feldspar (orthoclase), but a white or gray feldspar may be either K-feldspar or plagioclase. Plagioclase, however, has characteristic striations that may be visible on cleavage faces, especially if the crystals are several millimeters in size. Figure 1.10 shows a large fragment of a plagioclase crystal with striations on it. Striations on smaller plagioclase crystals are difficult but not impossible to detect.

Colors of Igneous Rocks

The mineral constituents of an igneous rock impart a characteristic color to it. Hence, rock color is used as a first-order approximation in establishing the general mineralogic composition of an igneous rock. As already pointed out, color is a relative and subjective property when modified only by the adjectives *light, intermediate,* or *dark*. All observers would agree that a white rock is "light-colored," a black rock is "dark-colored," and a rock with half of its constituent minerals white and half of them black is a rock of "intermediate color." Mineral constituents are not all black or white, however. Some are pink, gray, and other colors, a fact that adds to the difficulty encountered in igneous rock classification for the beginning student. Nevertheless, the terms *light, intermediate,* and *dark* are useful terms, especially in the classification of hand specimens with an aphanitic texture. For example, the rocks in figures 1.25, 1.26, and 1.33 are light-colored. The rock in figure 1.29 is intermediate in color, and those in figures 1.28 and 1.30 are dark-colored.

Igneous Rock Classification

Table 1.7 is a chart that shows the names of about twenty common igneous rocks and their corresponding textures and mineralogic compositions. This classification scheme is a simplified version of a more complex classification system in which the number of rock names is three to four times the number contained in table 1.7. Some simplification at the level of the beginning student is a pedagogical necessity, and if you should find a rock in your laboratory collection that does not fit easily into one of the categories shown in the table, it may be because the table has been so simplified. For example, in figure 1.27 is an example of a *pegmatite,* a very coarse grained igneous rock commonly associated with the margins of plutons. In general the mineralogic composition of pegmatites is granitic, but they may contain extremely large crystals of uncommon minerals such as spodumene, beryl, tourmaline, or topaz. This type of rock texture does not easily fit into a simplified rock classification table and thus has not been included in table 1.7.

Table 1.7   Classification and Identification Chart for Hand Specimens of Common Igneous Rocks

CLASSIFICATION	TEXTURE	ROCK NAME			
	Phaneritic (Coarse-grained)	GRANITE	DIORITE	GABBRO	PERIDOTITE
			("Granitic Rock")		
	Phaneritic With phenocrysts	GRANITE PORPHYRY	DIORITE PORPHYRY	GABRO PORPHYRY	Rocks in this area of composition and texture are either nonexistent or too rare to be considered at the elementary level of rock classification.
	Aphanitic (Fine-grained)	RHYOLITE	ANDESITE	BASALT	
			(Felsite)		
	Aphanitic With phenocrysts	RHYOLITE PORPHYRY	ANDESITE PORPHYRY	BASALT PORPHYRY	
			(Felsite Porphyry)		
	Vesicular	PUMICE	SCORIA		
	Glassy	OBSIDIAN			
	Pyroclastic	Rhyolite Tuff, Andesite Tuff, Basalt Tuff, Volcanic Breccia, Agglomerate			

*Muscovite and biotite are accessory minerals and are not essential to the classes of rocks containing them. Amphibole and pyroxene are accessory minerals where shown as a thin dashed line in the granite group.

In table 1.7 the lower left-hand side of the chart presents the textural categories while the color categories are displayed across the top. The upper part of the chart shows the mineral constituents of the various rock types. The width of the bar for each mineral is roughly proportional to the percentage of that particular mineral present. For example, quartz ranges from 0% to 25% as a mineral constituent, potassium feldspar (orthoclase) from 0% to 80%, plagioclase from 10% to 70%, olivine from 0% to 90%, amphibole from 0% to 25%, and pryoxene from 0% to 30%. The highest percentage value is equivalent to the widest part of the bar. These bars are used as estimates of percentages for the various mineral constituents, and not as rigorous constraints. This is in keeping with the fact that only in phaneritic rocks and phenocrysts can the percentage of minerals in a given hand specimen be approximated by visual inspection.

The classification of aphanitic, vesicular, and glassy rocks by macroscopic means is dependent on color alone. For this reason, the chart contains alternate names that can be used if a particular specimen does not fit neatly into one of the boxes on the chart. For example, an aphanitic rock that seems too dark to be a rhyolite but too light to be an andesite can be called a *felsite*. Similarly, a rock intermediate between a granite and a diorite can be labeled a *granitic rock*.

A group of igneous rocks that does not fit easily into the general classification scheme is the *pyroclastic rocks,* those that are accumulations of the material ejected from explosive type volcanoes. The lavas of these volcanoes are characterized by *high viscosity* (they do not flow easily) and high silica content. They are rhyolitic or andesitic in composition. Their mineral constituents are difficult to determine.

The volcanic ash generated from an eruption is known as *tuff* when it becomes consolidated into a rock. Light-colored tuff is called *rhyolite tuff* (fig. 1.35), and tuff of intermediate color is called *andesite tuff*. In some tuffs, small beadlike fragments of volcanic glass occur. These features are called *lapilli,* and the rock containing them is *lapilli tuff*.

A rock composed of the angular fragments from a volcanic eruption is a *volcanic breccia*. A volcanic rock composed of volcanic bombs and other rounded fragments is known as an *agglomerate*.

Exercise 2. Identification of Common Igneous Rocks

The identification and classification of igneous rocks is based on their texture and mineralogy. Color is also useful, but it usually is a reflection of the mineral composition of the rock.

A collection of igneous rock hand specimens will be provided to you by your instructor. Figures 1.25 through 1.40 show some examples of hand specimens of common igneous rocks. Although these may be helpful in identification, remember that the color of igneous rocks may vary considerably. For example, compare figures 1.25 and 1.26. Recognition of texture and mineralogy are the keys to the identification of igneous rocks in hand specimen.

1. Group the specimens into the textural categories described for you in the text. Note the variation in grain size for the phaneritic (coarse-grained) specimens. Separate out those specimens with diagnostic features such as phenocrysts or vesicles.

2. For the phaneritic (coarse-grained) specimens, identify the minerals present in each specimen.

3. Using table 1.7, proceed to identify and name each specimen in your collection. Note that you may have more than one specimen of a given rock type. If you identify several specimens as granite, try to identify accessory minerals in each so that the name you assign is more definitive than just "granite." For example, a granite containing biotite should be called a *biotite granite*.

4. Your laboratory instructor will advise you as to the procedure to be used to verify your identification.

Figure 1.25   Pink granite.

Figure 1.26   White granite.

Figure 1.27   Pegmatite.

Figure 1.28   Gabbro.

Figure 1.29   Diorite.

Figure 1.30   Basalt.

Figure 1.31   Vesicular basalt.

Figure 1.32   Basalt porphyry.

Figure 1.33 Rhyolite.

Figure 1.34 Rhyolite porphyry.

Figure 1.35 Rhyolite tuff.

Figure 1.36 Andesite porphyry.

Figure 1.37 Pumice.

Figure 1.38 Scoria.

Figure 1.39 Obsidian.

Figure 1.40 Peridotite.

Earth Materials

25

Sedimentary Rocks

Origin

Sedimentary rocks are derived from preexisting materials through the work of mechanical or chemical agencies under conditions normal at the surface of the earth, or they may be composed of accumulations of organic debris. Rock weathering on land produces rock and mineral *detritus* (fragments) that are transported by gravity (falling) or by wind, water, or ice and deposited elsewhere on the earth's surface as *clastic sediments*. Weathering also dissolves rock material and makes it available in solution to streams, rivers, and groundwater that transport it to lakes and oceans where it may be deposited as a chemical precipitate or an evaporite.

After sediment has been deposited, it may be compacted and cemented into a coherent mass or sedimentary rock. The process or processes by which soft sediment is transformed into rock is called *lithification*. Sedimentary rocks display various degrees of lithification. In this manual we will be concerned only with those sedimentary rocks that are sufficiently lithified to permit their being displayed and handled as coherent hand specimens.

Sedimentary rocks provide a number of clues as to their history of transportation and deposition. Some of these features may be present in the hand specimens that will be available in the laboratory. These *sedimentary structures* include such features as bedding, cross-bedding, ripple marks, graded bedding, and mud cracks, among others. In addition, some rock types are characteristic of specific environments of deposition.

Sedimentary rock classification is based on texture and mineralogic composition. Both features are related to the origin and lithification of the original sediment, but the origin cannot be inferred from a single hand specimen. Therefore, the classification scheme used will emphasize the physical features and mineralogy of the rock rather than its exact mode of origin.

Occurrence of Sedimentary Rocks

Sedimentary rocks are formed in a wide variety of sedimentary environments. Any places that sediments accumulate are sites of future sedimentary rocks. *Sedimentary environments* are grouped into three major categories: *continental, mixed continental,* and *marine*. Within each of these broad categories several subcategories exist. An abbreviated classification of sedimentary environments is given in table 1.8. This table is not intended to show all of the possible environments of deposition, but rather to illustrate the environments in which most of the sedimentary rock types you will study here might have originated.

Table 1.8. Simplified and Abbreviated Classification of Sedimentary Environments and Some of the Rock Types Produced in Each

SEDIMENTARY ENVIRONMENT	SEDIMENTARY ROCK TYPE
CONTINENTAL	
Desert	SANDSTONE
Glacial	TILLITE
River Beds	SANDSTONE CONGLOMERATE
River Floodplains	SILTSTONE
Alluvial Fans	ARKOSE, CONGLOMERATE, SANDSTONE
Lakes	SHALE, SILTSTONE
Swamps	LIGNITE, COAL
Caves, Hot Springs	TRAVERTINE
MIXED CONTINENTAL AND MARINE	
Littoral (between high and low tide)	SANDSTONE, COQUINA, CONGLOMERATE
Deltaic	SANDSTONE, SILTSTONE, SHALE
MARINE	
Neritic (low tide to edge of shelf)	SANDSTONE, ARKOSE, REEF LIMESTONE CALCARENITE
Bathyal (400 to 4000 m depth)	CHALK, ROCK SALT, ROCK GYPSUM, SHALE, LIMESTONE, GRAYWACKE
Abyssal (depths more than 4000 m)	DIATOMITE, SHALE

Because the earth's surface is dynamic and ever-changing, sedimentary environments in a given geographic location do not remain constant throughout geologic time. Areas that are now above sea level may have been covered by the sea at various times during the geologic past. Glaciers that existed in the geologic past have since disappeared. Lakes have formed or have dried up. As the depositional environment in a given locality changes through time, the kind of sediment deposited there changes in response to the new environmental conditions.

Sediments that accumulate in a particular sedimentary environment are generally deposited in layers that are horizontal or almost horizontal. The lateral continuity of these layers, or *beds,* reflects the areal extent and uniformity of the environment in which they were deposited. The thickness of a particular bed is a function of the *time* during which the depositional environment remained more or less constant and the *rate* at which the sediment accumulated. Separating the beds are *bedding planes,* which

Table 1.9. Simplified Particle Size Range, Clastic Sediment Name, and Associated Sedimentary Rock Type.

Particle Size Range	Sediment	Rock
Over 256 mm (10 in)	Boulder	Conglomerate (rounded fragments) or breccia (angular fragments)
2 to 256 mm (0.08 to 10 in)	Gravel	Conglomerate or breccia
1/16 to 2 mm (0.025 to 0.08 in)	Sand	Sandstone
1/256 to 1/16 mm (0.00025 to 0.025 in)	Silt	Siltstone*
Less than 1/256 mm (less than 0.00015 in)	Clay	Claystone*

*Both siltstone and claystone are also known as mudstone, commonly called *shale* if the rock shows a tendency to split on parallel planes.
From Carla W. Montgomery, *Fundamentals of Geology.* Copyright © 1989 Wm. C. Brown Publishers, Dubuque, Iowa. All Rights Reserved.
Reprinted by permission.

represent a slight or major change in the depositional history. These horizontal surfaces tend to be surfaces along which the sedimentary rocks break or separate. Rates of deposition vary widely, and a layer of shale 10 feet thick may represent a longer period of geologic time than a sandstone layer 100 feet thick.

When the environment of deposition changes in a particular region, the nature of the sediments accumulating there also changes. A bed of sandstone on top of a bed of shale reflects a change in sedimentary conditions from one in which clay was deposited to one in which sand was deposited. Both the underlying shale and the overlying sandstone represent the passage of geologic time.

Textures of Sedimentary Rocks

Texture is the appearance of a rock that results from the size, shape, and arrangement of the mineral grains in the rock.

Clastic Texture

Rocks with a *clastic* texture are derived from the detritus or rock waste that has been transported and deposited. (NOTE: In some cases, the terms *detrital* or *detritus* are used synonymously with clastic as a textural term. We have chosen to use detritus as the source, clastic as the resulting texture.)

The size of the individual particles is one of the chief means of distinguishing sedimentary rock types, and clastic particles are named according to their dimensions (i.e., their average grain diameters). Table 1.9 shows the names and sizes of sedimentary particles that will be useful in the classification of sedimentary rocks. This is a simplified version of a much more detailed classification system.

Particles larger than about ¼ mm can be distinguished with the naked or corrected eye. (Grains of ordinary table salt range in size from about ¼ to ½ mm in diameter. The dot in the letter "i" is about ½ mm in diameter, and a lowercase "l" is about 2½ mm high.) Sand grains smaller than ¼ mm and the larger silt grains can be distinguished with a hand lens, but the smaller grains of silt and clay can be distinguished only with a microscope. Be aware that the term *clay* is used here to define grain size rather than mineralogy, and take care to use the terms "clay mineral" and "clay size" to avoid confusion.

Sedimentary rocks composed mainly of silt or clay size particles are said to have a *dense* texture. The term *dense* is also a general textural term for any rock in which the individual mineral components are microscopic in size.

The shapes of the particles in the rock also influence texture. The mineral grains or rock fragments in a specimen with a clastic texture should be examined to determine whether they are *angular* or *rounded*.

Clastic texture is also influenced by the variation of particle size in the rock. Rocks with a predominant grain size are said to be *well sorted;* those with a wide variation in grain size are said to be *poorly sorted.*

Crystalline Texture

This texture is characteristic of a sedimentary rock composed of interlocking crystals. If the individual crystals are less than ¼ mm in diameter, the rock has a *dense texture* as far as macroscopic examination of a hand specimen is concerned.

Amorphous Texture

This is a very dense texture found in rocks composed of finely divided noncrystalline material deposited by chemical precipitation.

Oolitic Texture

This texture is formed by spheroidal particles less than 2 mm in diameter, called *oolites*. The oolites are usually composed of calcium carbonate or silica. They form by deposition of the material out of solution onto a nucleus (much like the formation of a hailstone), and they are cemented together into a coherent rock.

Bioclastic Texture

This texture is produced by the aggregation of fragments of organic remains, the most common of which are shell fragments or plant fragments. Rocks that contain fragments, molds, or casts of organic remains or *fossils* (e.g., shells, bones, teeth, leaves, seeds) or other recognizable evidence of past life (e.g., footprints, leafprints, worm burrows, etc.) are said to be *fossiliferous.* Fossils are commonly embedded in a matrix of sandstone, shale, limestone, or dolomite.

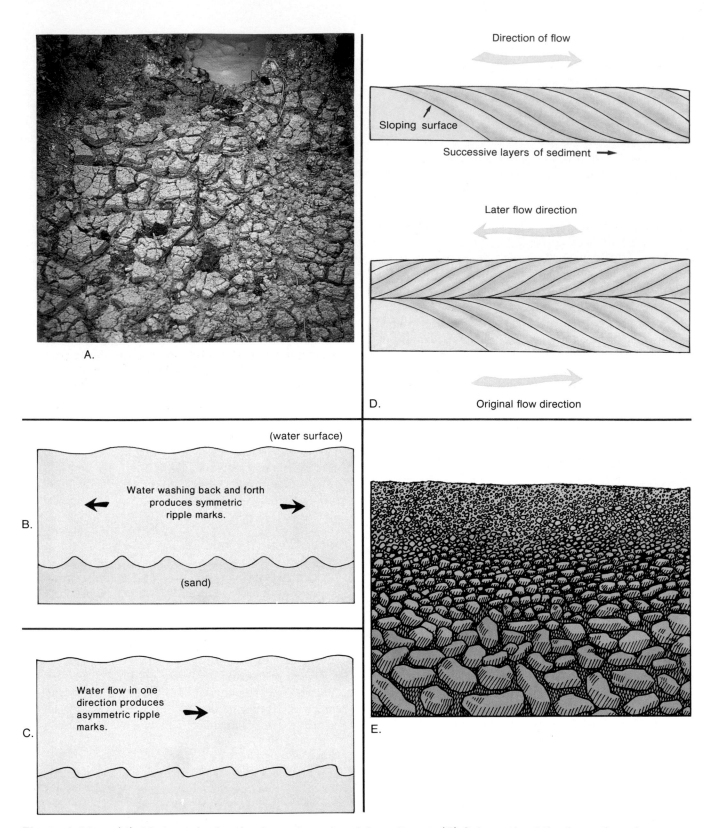

Figure 1.41 (*A*) Mud cracks forming in modern clay-rich sediment. (*B*) Schematic of the formation of oscillation ripple marks. (*C*) Schematic of the formation of current ripple marks. (*D*) Schematic example of cross-bedding formation. (*E*) Schematic of the formation of graded bedding. A–E From Carla W. Montgomery, *Physical Geology*, 2d ed. Copyright © 1990 Wm. C. Brown Publishers, Dubuque, Iowa. All rights reserved. Reprinted by permission.

Sedimentary Structures

Sedimentary rocks may contain features that are characteristic of their environment of deposition. These include *ripple marks, cross-bedding, graded bedding, mud cracks,* and *sole marks* among others. Even in hand specimen, many of these features may be evident and provide some indication of the sedimentary environment in which the rock formed. For example, the mud cracks found in rocks formed in exactly the same way as the mud cracks we see formed on dried mud today (fig. 1.41*A*). *Oscillation (symmetric) ripple marks* (fig. 1.41*B*) are characteristic of sediments deposited where there was a forward and backward movement of water such as one might find in a standing body of water affected by wave action. *Current (asymmetric) ripple marks* (fig. 1.41*C*) indicate that the sediment was deposited in running water. *Cross-bedding* is characteristically laid down at an angle to the horizontal, such as on the lee side of a sand dune, and while the major bedding is horizontal, there is a subset that is at an angle (fig. 1.41*D*). *Graded bedding* (fig. 1.41*E*) is the gradual vertical shift from coarse to fine clastic material in the same bed. *Sole marks* occur when there has been some disturbance of the top of a soft sediment that is then preserved when additional material is deposited. These markings are often tracks and trails of bottom-dwelling organisms.

These, plus other internal features of sedimentary rocks, can be most useful in deciphering the sedimentary environment and history of a given rock. If you take a field trip as part of your course, you will have an opportunity to see many of these features in the rocks you will observe.

Composition of Sedimentary Rocks

In hand specimen, the mineral constituents of sedimentary rocks are generally less varied than those in igneous rocks. The weathering of preexisting rocks involves chemical reactions that act upon the minerals within the rocks. Calcite, for example, may be dissolved and go into solution. In other minerals, these reactions result in a complete alteration of the mineral and often result in new minerals.

The stability of minerals at the surface of the earth varies greatly. Recall the reaction series discussed in the section on igneous rocks (see fig. 1.22). During the weathering process, those minerals that formed at high temperatures are the least stable at the surface of the earth and are most susceptible to chemical weathering. Figure 1.42 relates the Bowen Reaction Series to mineral stability. Thus, it is not surprising that olivine grains are relatively rare in sedimentary rocks and that quartz is

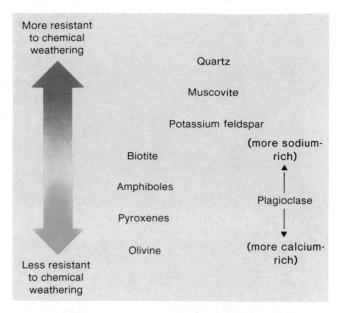

Figure 1.42 Susceptibility of minerals to chemical weathering is inversely related to Bowen's Reaction Series. From Carla W. Montgomery, *Physical Geology*, 2d ed. Copyright © 1990 Wm. C. Brown Publishers, Dubuque, Iowa. All rights reserved. Reprinted by permission.

common. The end products of the chemical weathering of most silicate minerals are quartz and clay minerals. A great many chemical reactions occur during the weathering, transportation, deposition, and lithification of sediments, all of which have some influence on the final composition of the rock.

Because some of the minerals that occur are often microscopic in size, they cannot always be identified easily. Tests for hardness and chemical composition are employed where visual inspection alone fails to identify the primary substance of which a sedimentary rock is composed.

The following materials are common in sedimentary rocks. Some of these are shown in the suite of sedimentary rocks in figures 1.43 through 1.50.

Silica

Silica, SiO_2, commonly occurs as quartz grains, either as a major or minor constituent of the rock or as minor amounts in other rocks. Quartz is the most ubiquitous of all minerals in sedimentary rocks. Other forms of silica include *chert,* a dense, cryptocrystalline form of quartz, and *diatomite,* a porous accumulation of the remains of siliceous plants of microscopic size.

Figure 1.43   Sandstone.

Figure 1.44   Arkose.

Figure 1.45   Conglomerate.

Figure 1.46   Shale.

Figure 1.47   Breccia.

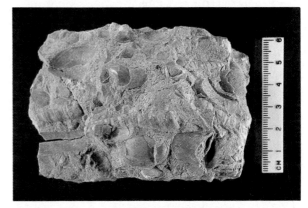

Figure 1.48   Fossiliferous limestone.

Figure 1.49   Chalk.

Figure 1.50   Coquina.

Carbonates

Calcite, $CaCO_3$, and dolomite, $CaMg(CO_3)_2$, are two common minerals that occur as the major constituents of limestone (calcite) and dolomite (or dolostone to distinguish it from the mineral), or as the cementing material in a wide variety of clastic sediments. A sedimentary rock containing calcite in any form will effervesce (fizz) strongly when a drop of cold dilute HCl is placed on it. Powdered dolomite will react weakly with the same cold dilute HCl. When applying this test it is important to distinguish whether it is the grains or crystals that are reacting with the acid, or whether it is the cement that is a carbonate.

The names of sedimentary rocks in which a carbonate mineral is present but is not a major constituent are prefixed by the term *calcareous,* which designates the presence of significant amounts of calcite or dolomite. For example, a rock made up of sand-sized quartz grains cemented with calcite would be called a *calcareous quartz sandstone.*

Calcite occurs in the crystalline form in *crystalline limestone.* Many shell fragments are composed of calcite, and the resulting rocks may range from *fossiliferous limestone* (fig. 1.48) to rocks called *coquina* composed of almost 100% shell fragments (fig. 1.50). Most oolites are composed of calcite. Thus, there is a wide range of rocks called limestone ranging from the crystalline variety to coquina.

Clay Minerals

This term refers to a group of *silicate* minerals that have layered atomic structures. *Clay minerals* are not to be confused with *clay-size particles,* which are smaller than 1/256 mm in diameter. All clay minerals occur as clay-size particles, but not all clay-size particles are clay minerals.

Although one clay mineral, kaolinite, is white, most clay minerals are green to gray and impart a dark color to the sedimentary rock in which they occur. Clay minerals are most common in dense fine-grained rocks such as *mudstone, shale* (distinguished by its *fissility,* the tendency to part in thin layers), and *graywacke.* Many limestones contain appreciable amounts of clay. The adjective used to describe a rock containing some clay is *argillaceous,* as for example an *argillaceous limestone.*

Evaporites

Minerals that belong to this group include gypsum, $CaSO_4 \cdot 2H_2O$, and halite, NaCl. Both are formed by chemical precipitation from an aqueous solution. Gypsum may occur as crystals in sedimentary rocks, but in this form it usually formed after the rock was deposited.

Rock Fragments

Some sedimentary rocks contain very coarse detrital constituents such as pebbles, cobbles, or even boulders. These coarse materials are usually rock fragments rather than single minerals. *Conglomerates* are clastic sedimentary rocks containing rounded pebbles or cobbles (fig. 1.45). If the coarse rock fragments are angular, the rock name is *breccia* (fig. 1.47).

Feldspars and Other Minerals

Compared to quartz, calcite, and clay minerals, feldspars do not occur in great abundance in sedimentary rocks. Under certain circumstances, however, some sedimentary rocks may contain significant amounts of feldspar. A feldspar-rich rock is called *arkose* (fig. 1.44). The corresponding adjective for rocks with some feldspar is *arkosic.* Feldspar also occurs in variable amounts in graywacke.

Organic Constituents

As noted above, organic remains may make up significant amounts of a given sedimentary rock. These materials may include shell fragments, teeth, bones, plant remains, or other organic debris.

Classification of Sedimentary Rocks

The great variety of sedimentary environments in nature and the gradations between them are responsible for the large diversity of sedimentary rock types. Because of this, a classification scheme encompassing all possible varieties would be unduly complex for the beginning student.

The threefold classification of sedimentary rocks presented in table 1.10 is highly simplified and is based on the major constituents of sedimentary rocks: *inorganic detrital material, organic detrital material,* and *inorganic chemical precipitates.* This arrangement is a logical and useful guide for an understanding of the similarities and differences in sedimentary rock types encountered by the beginning student.

The identification scheme for use with table 1.10 is presented as a part of Exercise 3.

Table 1.10 Classification and Identification Chart for Hand Specimens of Common Sedimentary Rocks

ORIGIN	TEXTURAL FEATURES AND PARTICLE SIZE	COMPOSITION AND/OR DIAGNOSTIC FEATURES	ROCK NAME
INORGANIC DETRITAL MATERIALS	Clastic. Pebbles and granules embedded in matrix of cemented sand grains.	Angular rock or mineral fragments.	BRECCIA
		Rounded rock or mineral fragments.	CONGLOMERATE
	Clastic. Coarse sand and granules.	Angular fragments of feldspar mixed with quartz and other mineral grains. Pink feldspar common.	ARKOSE
	Clastic. Sand-size particles.	Rounded to subrounded quartz grains. Color: white, buff, pink, brown, tan.	QUARTZ SANDSTONE
		Calcite and/or dolomite grains. Light colored.	CALCARENITE
	Clastic. Sand-size particles mixed with clay-size particles.	Quartz and other mineral grains mixed with clay. Color: dark gray to gray-green.	GRAYWACKE
	Clastic. Fine-grained. Silt and clay-size particles.	Mineral constituents not identifiable. Soft enough to be scratched with fingernail. Usually well stratified. Fissile (tendency to separate in thin layers). Color: variable.	SHALE
		Mineral constituents not identifiable. Soft enough to be scratched with fingernail. Massive (earthy). Color: variable.	MUDSTONE
INORGANIC CHEMICAL PRECIPITATES	Dense, crystalline or oolitic.	$CaCO_3$; effervesces freely with dilute HCl. May contain fossils (*fossiliferous*). Some varieties are *crystalline*. Some varieties are *oolitic*. Color: white, gray, black. Generally lacks stratification.	LIMESTONE
	Dense or crystalline.	$CaMg(CO_3)_2$; powder effervesces weakly with dilute HCl. May contain fossils (*fossiliferous*). Color variable, but commonly similar to limestones. Stratification generally absent in hand specimens.	DOLOMITE
	Dense, porous.	$CaCO_3$; effervesces freely with dilute HCl. Color variable. Contains irregular dark bands.	TRAVERTINE
	Dense (amorphous).	Scratches glass, conchoidal fracture. Color: black, white, gray.	CHERT
	Crystalline.	$CaSO_4 \cdot 2H_2O$; commonly can be scratched with fingernail. Color variable; commonly pink, buff, white.	ROCK GYPSUM
		NaCl. Salty taste. White to gray. Crystalline. May contain fine-grained impurities in bands or thin layers.	ROCK SALT
ORGANIC DETRITAL MATERIALS	Earthy (bioclastic).	$CaCO_3$; effervesces freely with dilute HCl; easily scratched with fingernail. Microscopic organisms. White color.	CHALK
		Soft. Resembles chalk but does not react with HCl. Commonly stratified. Gray to white. Microscopic siliceous plant remains.	DIATOMITE
	Bioclastic.	$CaCO_3$; calcareous shell fragments in a massive or crystalline matrix.	LIMESTONE
	Bioclastic.	$CaCO_3$; calcareous shell fragments cemented together.	COQUINA
	Fibrous (bioclastic).	Brown plant fibers. Soft, porous, low specific gravity.	PEAT
	Dense (bioclastic).	Brownish to brown-black. Harder than peat.	LIGNITE
	Dense (bioclastic).	Black, dull luster. Smudges fingers when handled.	BITUMINOUS COAL

Exercise 3. Identification of Common Sedimentary Rocks

A collection of sedimentary rocks will be provided in the laboratory by your instructor. Take time to examine the specimens so that you have some familiarity with the various types provided. In figures 1.43 through 1.50, examples of some sedimentary rocks are shown. Remember that there is a wide diversity of sedimentary rocks, and these examples may not exactly represent the rock types provided to you. As you complete the following parts of the exercise, keep a written record of the procedures you followed to determine the name for each specimen.

1. For each specimen, determine if (A) it is composed of mineral or rock fragments (inorganic detrital material), or (B) it is generally light colored and crystalline, oolitic, or dense (chemical precipitate), or (C) it is composed of fossil materials from animals or plants (organic detrital material).

2. For those specimens in group A, estimate the grain size and shape. Then determine the major constituent (i.e., quartz grains, rock fragments, etc.). Test with a drop of cold dilute HCl if you suspect the presence of carbonate. Using the appropriate part of table 1.10, determine the rock name for each specimen in this group.

3. For those specimens in group B, test with a drop of cold dilute HCl (remembering that dolomite will only react with cold dilute HCl when it is in a powdered form).

a) If there is a reaction, examine the specimen for other diagnostic features such as oolites, possibly some fossil remains, etc. Determine the rock name from table 1.10 for each specimen in this category.

b) If there is no reaction, test for hardness. Carefully apply a taste test if appropriate. Determine the rock name from table 1.10 for each specimen in this category.

4. For those specimens in group C, sort out the light-colored, fine-grained rocks.

a) Apply a drop of cold dilute HCl to the light-colored fine-grained specimens. Test for hardness. Examine them for other features. Determine the rock name from table 1.10 for each specimen in this category.

b) For the remaining specimens, the composition and other diagnostic features should allow you to determine the rock name from table 1.10 for each specimen in this category.

5. Where possible, apply the appropriate adjective to the sample; for example, fossiliferous limestone or calcareous quartz sandstone. Be as specific as possible. This will require special attention to *all* the features of the specimens.

6. Your laboratory instructor will advise you as to the procedure to be used to verify your identification.

Metamorphic Rocks

Once formed, all rocks are subject to processes of change that occur at the surface of the earth or within the crust of the earth. As we have seen, the processes with which we are familiar are those that take place at the surface of the earth and that combine to form sedimentary rocks. *Metamorphic* rocks are formed at varying depths within the crust when preexisting rocks are changed physically or chemically under conditions of high temperature, high pressure, or both. The process of *metamorphism*, literally a "change in form," takes place deep beneath the earth's surface and acts on all rocks, be they igneous, sedimentary, or metamorphic (refer to the rock cycle diagram, fig. 1.21).

As noted in the discussion of igneous rocks, the geothermal gradient is one source of heat for metamorphism. The other source is the heat from igneous plutons as they rise upward into country rock. Although extremely high temperatures may occur, metamorphism always occurs *below* the melting point of the rocks. If melting occurs, the process is considered igneous.

Pressure derives from two sources. The first is the *confining pressure* on the deeply buried rocks resulting from the weight of the overlying rocks. Confining pressure is equal in all directions and is related to the depth of burial. The second is *directed pressure*, which as the name suggests is greater in one particular direction. An example of directed pressure is that associated with the process of mountain building.

The process of metamorphism is aided by the presence of fluids in the rocks. While it is possible for chemical elements to migrate through the country rock without fluids being present, their movement is greatly facilitated by fluids.

The mineralogy and texture of metamorphic rocks provide some insight into the temperature and pressure environment in which they were formed. *Metamorphic facies* are defined by the typical assemblage of minerals found in the rocks formed at specific combinations of temperature and pressure. *Metamorphic grade* refers to the intensity of metamorphism in a given rock.

Metamorphic Environments

The metamorphism associated with the intrusion of a magma into the country rock (sometimes referred to as the host rocks) results in *contact metamorphism*. Both the heat and the chemical constituents which emanate from the magma produce mineralogical changes in the host rocks. Contact metamorphism is most intense at or near the zone of contact between the magma and the host rock, and the effects of contact metamorphism progressively diminish in a direction away from the contact zone. This *aureole* or halo of metamorphism surrounding the magma is characterized by the formation of high-temperature minerals close to the contact zone and progressively lower temperature minerals as the distance from the contact zone increases. Normally there is no evidence of the reorientation of the minerals. The effect of migration of materials from the magma into the host rock, *metasomatism*, may or may not occur. The size of the aureole depends on the temperature and size of the pluton, mineralogy of the host rock, and the presence or absence of fluids.

By far the largest volume of metamorphic rocks is produced by those processes associated with *regional metamorphism*. Regional metamorphism is a result of large-scale *tectonic* or *mountain-building* events involving movement and deformation of the earth's crust on a regional scale. During periods of mountain building, large segments of crustal rocks are deformed. These deformed areas occur in zones or belts hundreds of miles wide and thousands of miles long (the Appalachian Mountain belt is an example). Rocks in these belts are subjected to stretching (tensional) and squeezing (compressional) stresses that cause physical and mineralogical changes in the rock. They are subjected to extreme stresses, and rocks at great depth may deform as a plastic rather than as a brittle solid. This accounts for the fact that many rocks deformed under conditions of regional metamorphism have a texture, called *foliation*, that is characterized by a parallel arrangement of platy minerals such as the micas or the common orientation of the long axis of such minerals as hornblende.

A third but less important environment in which metamorphism can occur is that associated with *fault zones* such as the San Andreas fault in California (discussed in Exercise 24). In this environment high pressure and low temperature are present, and the materials commonly formed are composed of broken and distorted fragments of the rocks on either side of the fault zone and minerals that form only at low temperature and high pressure. If coarse-grained, these rocks are known as *fault breccia*, if fine-grained as *mylonite*.

Metamorphic Facies and Grade

Contact and regional metamorphism produce differing degrees of change in preexisting rocks. As briefly discussed above, the aureole associated with contact metamorphism is defined by a decrease in degree of metamorphism as the distance from the pluton increases. The same is true of regional metamorphism; rocks on the outer margins of a mountain belt that has been subjected to a single period of deformation may be only slightly metamorphosed. The metamorphic rocks in the center of the belt may be so deformed that the texture and mineralogy of the original or parent rock have been obliterated.

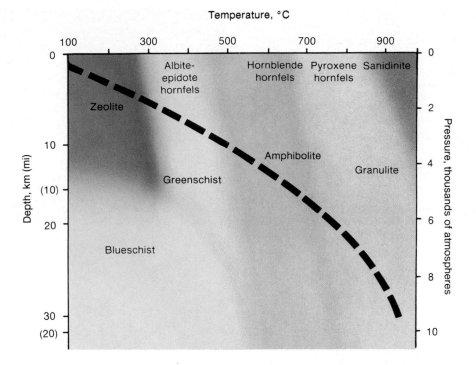

Figure 1.51 Metamorphic facies as functions of temperature and pressure (depth). Dashed curve is average continental geothermal gradient. Adapted from Carla W. Montgomery, *Physical Geology*, 2d ed. Copyright © 1990 Wm. C. Brown Publishers, Dubuque, Iowa. All rights reserved. Reprinted by permission.

The study of metamorphic rocks has led to the identification of *metamorphic facies:* the recognition of an assemblage of minerals in the rocks that formed during metamorphism under specific environmental conditions of temperature and pressure. The assemblages formed depend also on the mineralogy of the parent rock, but the same parent rock will yield the same assemblages at the same combination of temperature, pressure, and fluids. The commonly recognized metamorphic facies are named for characteristic minerals or characteristic rock types associated with them (fig. 1.51).

If we consider the effects of contact metamorphism on a single rock type, the resulting facies are relatively simple because temperature is the single important variable. The facies recognized are named for specific minerals: *sanidinite,* the high-temperature facies, is characterized by the presence of the mineral sanidine, a variety of orthoclase, grading outward from the heat source through the *hornfels* facies to the *zeolite* or low-temperature facies.

The facies relationships associated with regional metamorphism are more complex due to the fact that not only are there two variables, both temperature and pressure, but the size of the area affected is such that there is a wide variety of rock types involved. Not all facies may be present in a given region. Figure 1.51 is a schematic diagram of the temperature and pressure relationships of the commonly recognized metamorphic facies.

Metamorphic grade is the measure of the intensity of metamorphism to which a given rock has been subjected.

Table 1.11. Simplified table relating rock type, metamorphic grade, and index minerals.

ROCK TYPE	METAMORPHIC GRADE	INDEX MINERAL
Slate	Low	
Phyllite	Low to intermediate	Chlorite
Schist	Intermediate to high	Garnet
Gneiss	High	Sillimanite

Low-grade metamorphism occurs in the marginal area, *high-grade* metamorphism occurs where the effects of temperature and pressure have been most intense, and *intermediate-grade* metamorphism lies in between. Table 1.11 is a simplified schematic showing the relationship between metamorphic grade, rock type, and the *index minerals* associated with each. Index minerals are those minerals that are stable over a specific range of temperature and pressure conditions and that are useful in determining metamorphic grade (fig. 1.52).

The metamorphism of shale, a common sedimentary rock, provides a simple way of illustrating metamorphic grade. The sequence of increasing metamorphic grade shows the gradual change from shale to slate to phyllite to schist to gneiss. In this simplified presentation, slate represents low-grade metamorphism, phyllite low to intermediate, schist intermediate to high, and gneiss high-grade metamorphism. As metamorphic grade increases, there are corresponding mineralogic and textural changes that occur within the rocks.

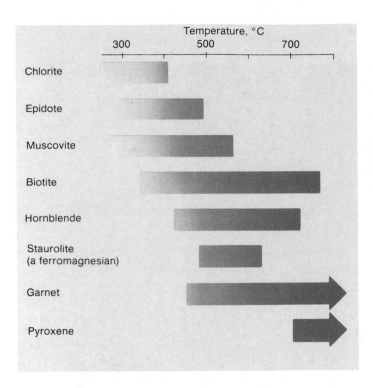

Temperature, °C

Figure 1.52 Approximate temperature ranges over which some representative index minerals are stable. From Carla W. Montgomery, *Physical Geology*, 2d ed. Copyright © 1990 Wm. C. Brown Publishers, Dubuque, Iowa. All rights reserved. Reprinted by permission.

Texture and Composition of Metamorphic Rocks

Metamorphic textures consist of two main types, *foliated* and *nonfoliated.* The mineral constituents of foliated metamorphic rocks are oriented in a parallel or subparallel arrangement. Foliated metamorphic rocks are generally associated with regional metamorphism. The nonfoliated rocks exhibit no preferred orientation of mineral grains and most commonly reflect the effects of contact metamorphism.

Foliated Textures

Four kinds of foliated textures are recognized. In order of increasing metamorphic grade, these are *slaty, phyllitic, schistose,* and *gneissic* (pronounced nice-ick).

Slaty Texture This texture is caused by the parallel orientation of microscopic grains. The name for the rock with this texture is *slate* (fig. 1.53), and the rock is characterized by a tendency to separate along parallel planes. This feature is a property known as *slaty cleavage.* (Slaty cleavage or rock cleavage is not to be confused with cleavage in a mineral, which is related to the internal atomic structure of the mineral.)

Phyllitic Texture This texture is formed by the parallel arrangement of platy minerals, usually micas, that are barely macroscopic (visible to the naked or corrected

Figure 1.53 Slate, showing characteristic slaty cleavage.

eye). The parallelism is often wavy, or crenulated. The predominance of micaceous minerals imparts a sheen to the hand specimens. A rock with a phyllitic texture is called a *phyllite.*

Schistose Texture This is a foliated texture resulting from the subparallel to parallel orientation of platy minerals such as chlorite or micas. Other common minerals present are quartz and amphiboles. A schistose texture lies between the parallel platy appearance of phyllite and the distinct banding of gneissic texture. The average grain size of the minerals is generally smaller than in a gneiss (see below). A rock with schistose texture is called a *schist* (figs. 1.54 and 1.55).

Gneissic Texture This is a coarsely foliated texture in which the minerals have been segregated into discontinuous bands, each of which is dominated by one or two minerals. These bands range in thickness from 1 mm to several centimeters. The individual mineral grains are macroscopic and impart a striped appearance to a hand specimen (figs. 1.56 and 1.57). Light-colored bands commonly contain quartz and feldspar, and the dark bands are commonly composed of hornblende and biotite. Accessory minerals are common and are useful in applying specific names to these rocks. A rock with a gneissic texture is called a *gneiss.*

Nonfoliated Texture

Metamorphic rocks with no visible preferred orientation of mineral grains have a nonfoliated texture. Nonfoliated rocks commonly contain equidimensional grains of a single mineral such as quartz, calcite, or dolomite. Examples of such rocks are *quartzite* (fig. 1.58), formed from a quartz sandstone, and *marble* (fig. 1.59), formed from a limestone or dolomite (dolostone). Conglomerate that has been metamorphosed may retain the original textural characteristics of the parent rock, including the outlines and colors of the larger grain sizes such as granules and pebbles. However, because metamorphism has caused recrystallization of the matrix, the metamorphosed

Figure 1.54 Garnetiferous schist.

Figure 1.55 Biotite schist.

Figure 1.56 Gneiss.

Figure 1.57 Gneiss.

Figure 1.58 Quartzite.

Figure 1.59 Pink marble.

Figure 1.60 Metaconglomerate.

Figure 1.61 Anthracite coal.

conglomerate is called *metaconglomerate* (fig. 1.60). In some cases the metamorphism has deformed the shape of the granules or pebbles; in this case the rock is called a *stretched pebble conglomerate.*

Quartzite and metamorphosed conglomerate can be distinguished from their sedimentary equivalents by the fact that they break *across* the quartz grains, not around them. Marble has a crystalline appearance and generally has larger mineral grains than its sedimentary equivalents.

A fine-grained (dense textured), nonfoliated rock usually of contact metamorphic origin is *hornfels.* Hornfels has a nondescript appearance because it is usually some medium to dark shade of gray, is lacking in any structural characteristics, and contains few if any recognizable minerals in hand specimen.

The metamorphic equivalent of bituminous coal is *anthracite coal* (fig. 1.61), and a sequence similar to that presented for the metamorphism of shale would show the change from peat to lignite to bituminous coal to anthracite coal and, under extreme conditions, to the mineral graphite.

The Naming of Metamorphic Rocks

Foliated metamorphic rocks are named according to their texture (e.g., slate, phyllite, schist, or slate). In addition to the root name, the name or names of the dominant or distinctive (but not abundant) minerals may be added as descriptors. In some rocks, mineral crystals have formed that are much larger than the matrix in which they are contained. These are called *porphyroblasts,* and recognition of their mineralogy aids in the detailed description of the rock in which they are contained. For example, a schist with recognizable garnet grains (porphyroblasts) would be called a *garnet schist* (some would use the term "garnetiferous" as the descriptor), and the exact textural description would be a rock with a porphyroblastic schistose texture. Other examples of rock names with mineral descriptors are quartz-hornblende gneiss, chlorite schist, biotite-garnet schist, garnet gneiss, amphibole schist, and granite gneiss.

The names of slate are commonly modified by a color name, as in green slate, red slate, or black slate.

In the case of nonfoliated rocks, a color prefix is also commonplace in the naming of quartzite or marble. Such names as white marble, pink marble, or variegated marble

Table 1.12. Classification and Identification Table for Hand Specimens of Common Metamorphic Rocks

TEXTURE	DIAGNOSTIC FEATURES	ROCK NAME
FOLIATED	Gneissic texture. Coarse grained. Foliation present as macroscopic grains arranged in alternating light and dark bands. Abundant quartz and feldspar in light-colored bands. Dark bands may contain hornblende, augite, garnet, or biotite.	GNEISS
	Schistose texture. Medium to fine grained. Common minerals are chlorite, biotite, muscovite, garnet, and dark elongate silicate minerals. Feldspars commonly absent. Recognizable minerals used as part of rock name. Porphyroblasts common.	SCHIST
	Phyllitic texture. Fine grained to dense. Micaceous minerals are dominant. Has a sparkling appearance.	PHYLLITE
	Slaty texture (slaty cleavage apparent). Dense, microscopic grains. Color variable; black, and dark gray common. Also occurs in green, dark red, and dark purple colors.	SLATE
NONFOLIATED	Texture of conglomerate but breaks across coarse grains as easily as around them. Granules or pebbles are commonly granitic or jasper, chert, quartz, or quartzite. If pebbles are deformed, called *stretched pebble conglomerate.*	METACON-GLOMERATE
	Crystalline. Hard (scratches glass). Breaks across grains as easily as around them. Color variable; white, pink, buff, brown, red, purple.	QUARTZITE
	Dense, dark colored; various shades of gray, gray-green, to nearly black.	HORNFELS
	Crystalline. Composed of calcite or dolomite. Color variable; white, pink, gray, among others. Fossils in some varieties.	MARBLE
	Black, shiny luster. Conchoidal fracture.	ANTHRACITE COAL

(meaning a marble containing streaks of several different colors) are commonplace.

Classification of Metamorphic Rocks

The classification of metamorphic rocks is based on texture and composition, the basic distinction being between foliated and unfoliated textures (table 1.12). *The rock names given are only the root names,* because the chart would become too large and cumbersome if all possible varieties of gneiss, schist, marble, and quartzite were listed. You are encouraged to utilize as many descriptors as applicable in naming metamorphic rocks.

Exercise 4. Identification of Common Metamorphic Rocks

A group of metamorphic rock hand specimens will be provided to you in the laboratory. Take time to review the types of foliated textures described on the previous pages. The rocks depicted in figures 1.54–1.61 are examples of several types of metamorphic rocks and may assist you in identification. Remember that there are a wide variety of types, textures, and colors of metamorphic rocks, and your collection may include specimens that do not appear in these figures.

1. Divide your hand specimens into the two major textural categories—foliated and nonfoliated.

2. Arrange the foliated specimens in sequence from coarse grained to fine grained or dense.
   a) Work first with the coarse-grained rocks to apply a rock name. Examine each specimen carefully to identify specific minerals that are present either in abundance or as porphyroblasts. Utilize these when applying a specific rock name to the specimens of gneissic or schistose texture. Some examples that may be in your collection are:

   GNEISS:
   Biotite gneiss
   Biotite hornblende gneiss
   Quartz feldspar gneiss
   (Others possible)
   SCHIST:
   Garnet schist
   Muscovite schist
   Biotite schist
   Tourmaline mica schist

   Staurolite schist
   Hornblende schist
   Garnet mica schist
   (Others possible)

   b) The fine-grained to dense rocks with foliation should be examined on the basis of texture and then color. Descriptors are usually not possible with phyllite, and the slates are generally described by color. In some cases, they may contain enough calcium carbonate to react with cold dilute HCl and would be called "calcareous shale."
   c) Identify and apply rock names to each of the foliated specimens. For each, give an indication of the metamorphic grade represented.

3. Work next with the nonfoliated specimens. Remember the following:
   a) Marble is composed of calcite or dolomite, hence it is softer than glass and reacts with cold dilute HCl in the same way that limestone or dolomite (dolostone) does.
   b) Quartzite and metaconglomerate are rich in silica (quartz) and scratch glass.
   c) Graphite as a rock has the same characteristics as the mineral graphite.
   d) Identify and apply rock names to all specimens. Where possible use descriptors such as color, fossil content, shape of pebbles, etc.

4. Your laboratory instructor will advise you as to the procedure to be used to verify your identification.

The Geologic Column and Relative Geologic Time

Background

An understanding of the relationships between rock units and their relative ages is essential in the unraveling of the geologic history of a given area. The relative ages of sedimentary rocks are determined by the application of two basic geologic principles. The first is the *law of original horizontality,* which states that sediments deposited in water are laid down in strata that are horizontal or nearly horizontal. The second is the *law of superposition,* which states that in any undisturbed sequence of sedimentary rocks, the layer at the bottom of the sequence is older than the layer at the top of the sequence.

Through the application of the principles used to establish the relative ages of sedimentary strata around the world, a geologic time scale for all of earth history has been pieced together. The geologic time scale as used in North America is shown in table 1.13. The table is arranged with the oldest geologic ages at the bottom and the youngest at the top.

All areas of the earth's surface were not sites of deposition throughout all of geologic time. Some areas, especially mountain regions, contained preexisting rocks from which sedimentary materials were produced by natural decay and physical breakdown. These sediments were eroded and transported to sites of deposition by various geologic agents such as running water, glaciers, and wind.

A depositional site may change over geologic time to a site where erosion is taking place. Hence, in any given geographic locality, all of geologic time is not represented by a continuous sequence of strata. Many gaps in the sedimentary record occurred in one locality or another across the face of the earth during the passage of geologic time. Therefore, in order to construct a more complete geologic history of a particular area, fragments of that history recorded in rocks from different localities must be pieced together.

The most thoroughly documented part of the geologic time scale, sometimes referred to as the *geologic column,* comes from the geologic strata deposited during the last

Table 1.13. Geologic Time Scale as Used in North America. (Absolute ages based on Geological Society of America, The Decade of North American Geology 1983 Time Scale.)

ERA	PERIOD	EPOCH	MAP SYMBOL	COMMON MAP COLOR
CENOZOIC	Quaternary	Holocene Pleistocene	Q Q	Various shades of gray and yellow
	— 1.6 million years —			
	Tertiary	Pliocene Miocene Oligocene Eocene Paleocene	Pl or Tpl M or Tm Φ or To E or Te Tp	Various shades of orange, yellow-orange, and yellow
	— 66 million years —			
MESOZOIC	Cretaceous Jurassic Triassic		K J Ŧ	Various shades of green Various shades of blue-green Various shades of blue
	— 245 million years —			
PALEOZOIC	Permian Pennsylvanian Mississippian Devonian Silurian Ordovician Cambrian		P or Cpm IP or Cp M or Cm D S O €	Commonly blue, green, purple, pink, lavender, purple-gray Various shades of purple, pink, lavender, tan, brown, red-brown, red
	— 570 million years —			
PRECAMBRIAN			p€	No standard color

600 million years. This segment of geologic time has been divided into subunits called eras, periods, and epochs. The three major eras in order of decreasing age are the Paleozoic, Mesozoic, and Cenozoic. Each of these is divided into periods, and the periods are further divided into epochs. The names of the eras are based generally on the fossils contained in strata formed during those eras. *A fossil is any evidence of past life such as bones, shells, leaf imprints, and the like.* Paleozoic means "early life," Mesozoic means "middle life," and Cenozoic means "recent life." The names of the periods and epochs are based on strata originally studied in Europe during the eighteenth and nineteenth centuries; hence, the names are chiefly European in origin.

The time units of the geologic column are not of equal duration as can be seen from the absolute ages of the time boundaries between the eras in table 1.13. The geologic column as it existed in the early part of the twentieth century was also subdivided by absolute ages, but these dates lacked precision because they were based on inaccurate assumptions. The absolute ages in table 1.13 are based on age determinations on rocks containing radioactive materials that decay at a constant rate. By careful measurement of the components produced by radioactive decay, rocks can be dated with a great deal of precision.

Precambrian rocks are those that were formed prior to the beginning of the Cambrian period. For the sake of simplicity, the Precambrian is not divided into subdivisions in table 1.13. Generally, Precambrian rocks are metamorphic and igneous in type, and the field relationships are extremely complex in many areas where they occur.

The geologic time scale is introduced here primarily to provide a broader context in which the development of simple geologic columns representing very small segments of geologic time can be placed. Other information in table 1.13, such as the abbreviations of geologic time units and standard colors used on geologic maps, is for general information only, and has no application here. However, in Part 4 where geologic maps are considered in detail, the use of these abbreviations and map colors will become apparent.

We will now consider the means by which a geologic column for a given locality is constructed.

Constructing a Geologic Column

Figure 1.62 shows a series of rock strata in a *geologic cross-section* with a corresponding *geologic column* to the right. The geologic column is constructed by applying the law of superposition to the cross-section. It is apparent that the limestone in figure 1.62 is the oldest, the shale is younger than the limestone, and the sandstone is the youngest of the three formations. The process of sedimentation was continuous during the time it took for these

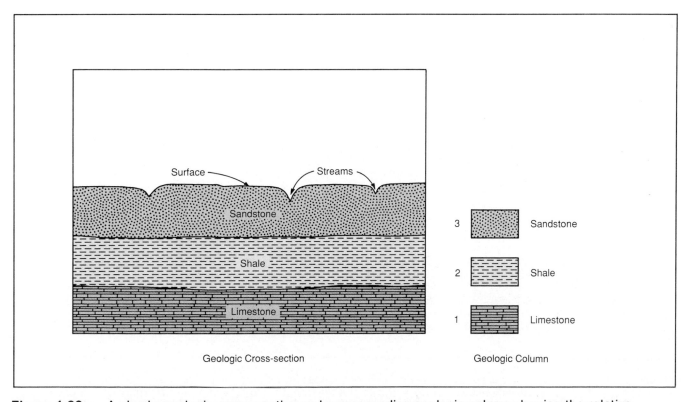

Figure 1.62   A simple geologic cross section and corresponding geologic column showing the relative ages of the three strata with the oldest (1) at the bottom and the youngest (3) at the top. Construction of the geologic column is based on the law of superposition.

three layers to be deposited; only the depositional environment changed. Figure 1.63 shows the symbols used to portray various rock types on geologic cross-sections and geologic columns.

Unconformities

From the information available in figure 1.62, one can conclude that some time after the sandstone was formed, the marine environment in which it formed changed to an environment of erosion or nondeposition. One cannot deduce from figure 1.62 whether additional strata were laid on top of the sandstone. All that can be said is that a depositional environment gave way to an erosional one some time after the sandstone was formed. Now, if the sea that once covered the area underlain by the three strata in figure 1.62 invaded the area again, and another sequence of strata was deposited, the situation would be depicted as in figure 1.64. The old erosional surface that is now buried beneath the younger strata is called an *unconformity* or a *surface of erosion*. It constitutes an unknown amount of geologic time that elapsed between the cessation of deposition of the older sequence of limestone-shale-sandstone and the upper sequence of conglomerate-arkose-siltstone.

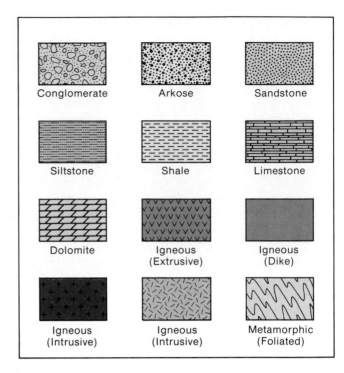

Figure 1.63 Symbols used in geologic columns and geologic cross sections. The colors are used to differentiate rock types only and bear no relationship to the colors used on geologic maps listed in table 1.13.

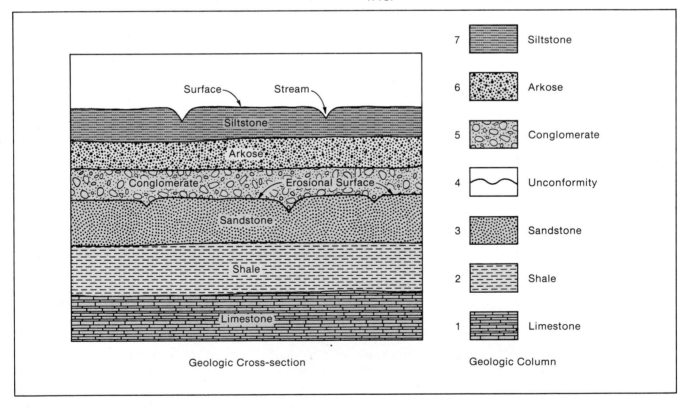

Figure 1.64 Geologic cross section showing two sequences of sedimentary strata separated by an unconformity. The geologic column at the right shows the sedimentary layers and unconformity of the cross section arranged in chronological order with the oldest at the bottom and the youngest at the top. The geologic time encompassed by the geologic column cannot be determined because only the relative ages of the rock layers can be deduced from the geologic cross-section.

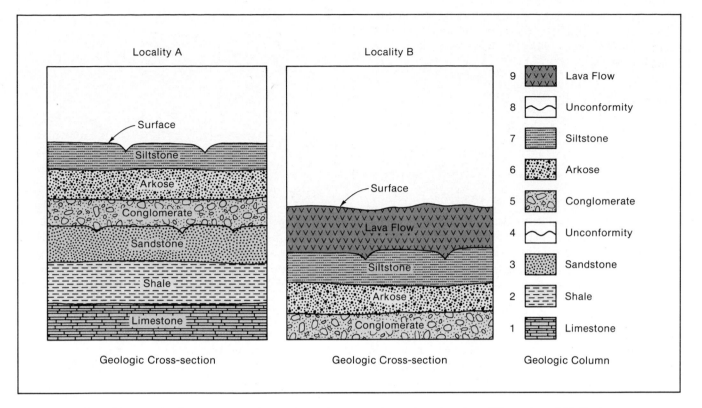

Figure 1.65 The geologic cross sections shown here are based on two localities, A and B, that are separated by a few miles in which no outcrops exist. By combining the stratigraphy of the two localities, a geologic column using information from both can be constructed. The geologic column is based on the assumption that the conglomerate-arkose-siltstone sequences in both localities are of the same geologic age, in which case they are said to be correlated. Correlation in this case is based on a similar stratigraphic sequence whose constituent beds have lithologic similarities.

In figure 1.64, the law of superposition still applies in constructing the geologic column. The relative ages of all six formations and the unconformity are shown by the appropriate numbers next to the boxes in the geologic column. The unconformity represents a break of unknown geologic duration in the sedimentary record for this particular locality. It is assumed that this unconformity represents a period of erosion, but the actual length of time in years represented by the unconformity cannot be determined from the information in figure 1.64. All that can be done is to place it in its relative chronological position in the geologic column.

Correlation

In figure 1.64, the two periods of sedimentation and the unconformity between them all represent the passage of geologic time in the same geographic locality. A geologic column constructed therefrom is based only on the rocks that crop out in that particular area. However, by tracing certain formations from this locality to others near by, it may be possible to extend the geologic column to include rocks older or younger.

As an example, consider figure 1.65 in which two geologic cross-sections are shown from two locations separated by a few miles. Locality A contains a sedimentary sequence with six lithologically distinct formations. At locality B, only the upper sedimentary sequence crops out; the lower one is out of sight, presumably beneath the surface, and hence not observable. At locality B, a lava flow lies on top of a siltstone. If, in fact, the siltstone at locality A and the siltstone at B are one and the same formation, they are said to be *correlated*. This being so, it is then possible to construct a geologic column from the stratigraphic information at both localities A and B as shown to the right in figure 1.65. The lava is the youngest formation in the column, and two unconformities are present, one between the siltstone and the lava flow, and one between the conglomerate and the sandstone.

Exercise 5. Building a Geologic Column from a Geologic Cross Section

Figure 1.66 is a schematic geologic cross-section based on the mapping of different outcrops from several localities in a geographic area covering about 10 square miles. Formations shown by the same color and symbols are the same age. Study the cross-section and complete the work called for in questions 1 and 2.

1. Construct a geologic column from this cross-section using the numbered lines at the right of figure 1.66 to record your interpretation. Indicate the name of a rock formation (e.g., shale, stock, alluvium) or an unconformity in its appropriate position in the column as determined from the adjacent geologic cross-section. Igneous rock masses of the same age should be listed on the same line. Remember that the oldest unit is listed at the bottom.

2. Re-examine figure 1.66. In this simplified geologic cross-section, no evidence of metamorphic activity has been shown. Recall the earlier discussion of types of metamorphism, metamorphic grade, and the classification of metamorphic rocks. Had metamorphic rocks been included, indicate what types you would expect to find, and why.

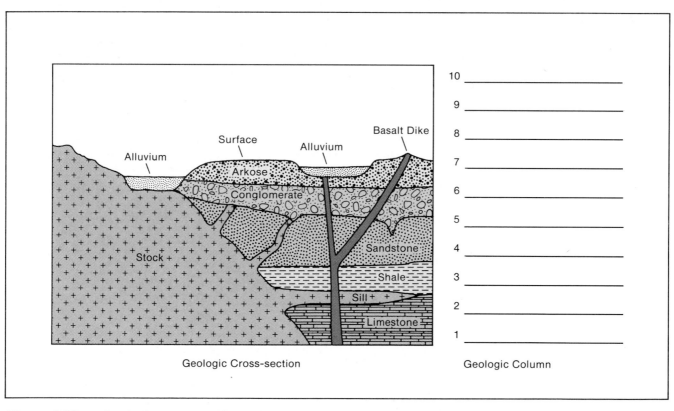

Figure 1.66 Geologic cross section and uncompleted geologic column. The names of the rock units and any unconformities are to be listed in the appropriate boxes of the geologic column as called for in Exercise 5.

Exercise 6 Relative Ages of Igneous Rocks Intruded into Country Rock

DUNCAN LAKE AERIAL PHOTOGRAPH, Canada

This photograph (fig. 1.67) was taken from an airplane with a camera whose lens was pointed vertically downward toward the surface of the earth. The picture covers an area of about 3 square miles (2 miles in the east-west direction and 1.5 miles in the north-south direction). The black areas are lakes. Because there are few trees and little vegetation in this part of northern Canada, the rocks are well exposed, or put in geologic terms, there are many outcrops. Fieldwork in the area reveals three rock types, each of which appears as a distinct shade of gray on the photograph. The very light gray areas are granite. The medium gray represents foliated metamorphic rocks, and basalt dikes occur in long narrow outcrop patterns. (When looking at the photo, orient it so that the north arrow is in the upper right-hand corner.)

Figure 1.68A is a schematic geologic cross section drawn from figure 1.67. Using both figure 1.67 and figure 1.68A, answer the following:

1. Which of the three kinds of rock is the country rock?
2. Does the granite show a discordant or concordant relationship with the country rock? (Examine the small patches of granite in the northeast corner of the photograph for clues. The trend of the country rock in this area is revealed by alternating linear bands of medium gray and somewhat darker gray areas.)

3. What are the modes of occurrence of the granite?

4. Are the basalt dikes younger or older than the granite?

5. List the three rock types by name at the right of the numbered boxes in figure 1.68B with the oldest in box 1 and the youngest in box 3.

NORTH

1200'

0'

Figure 1.67 Aerial photograph of the Duncan Lake area, Northwest Territories, Canada. (Courtesy of the Royal Canadian Air Force.)

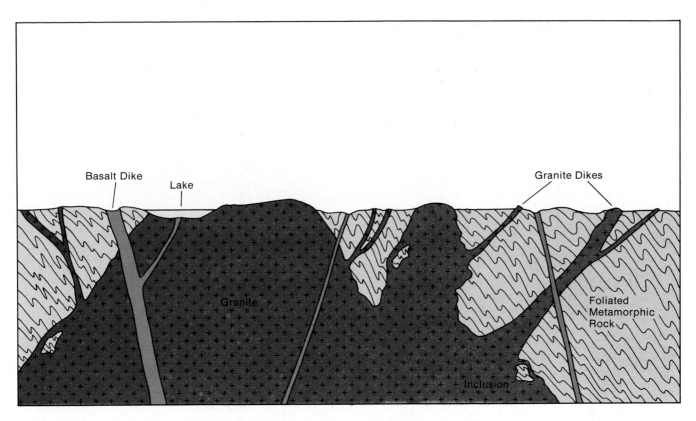

A. Generalized geologic cross-section of Duncan Lake aerial photograph of figure 1.67.

B. Geologic column to be completed following instructions in Exercise 6, question 5.

Figure 1.68 Diagrams for use in Exercise 6.

Topographic Maps, Aerial Photographs, and Other Imagery from Remote Sensing

Background

The earth is the laboratory of geologists. They are interested not only in the materials of which the earth is made, but also in the configuration of its surface. Two important tools are used by geologists: maps, and ordinary photographs supplemented by other remote sensing imagery.

A *map* is a representation of part of the earth's surface. Maps that show only the horizontal distribution of earth features and man-made structures are of limited use to the geologist because geologic phenomena are three-dimensional. Therefore, a map that portrays the earth's surface in three dimensions is of particular value to geologists. Maps that meet this requirement are topographic maps. The first section in Part 2 of this manual deals with topographic maps.

Remote sensing is a process whereby the image of a feature is recorded by a camera or other device and reproduced in one form or another as a "picture" of the feature. One of the oldest tools of remote sensing is the camera, which produces images on a photosensitive film. When developed by chemical means, the so-called black-and-white photographs or true-color photographs are the results.

The space age saw the introduction of many other kinds of remote sensing devices that produce new kinds of images such as "radar photographs," false color images, and a variety of other "pictures." These are useful not only to geologists but also to geographers, ecologists, foresters, soil scientists, meteorologists, and the like. The radar picture or image, for example, is made by recording the radiation from earth features in a way that can be resolved into a photolike picture. Radar images from earth-orbiting satellites are used extensively to show cloud cover on televised weather reports and forecasts over most commercial television stations. Radar images made from aircraft are useful in the study of many geologic phenomena.

False color photos or images are made with a device that records infrared radiation from the earth. The resulting image shows earth features in colors that are different from the true colors. For example, vegetation shows up as red instead of green on the false color images. False color pictures generally enhance the differences in earth features due to variations in vegetation, soil, water, and rock types.

Photographs or other types of images made by cameras or other sensing devices installed in airplanes, manned spacecraft, or satellites are thus another kind of map that provide geologists with useful tools for analyzing and interpreting the components of a given landscape on earth as well as on the moon and on other planets in the solar system.

The second section of Part 2 introduces students to aerial photographs and false color images. (A radar map will be introduced in Part 4.) For those interested in pursuing the subject of remote sensing further, attention is directed to the list of references at the end of Part 2.

The goal of Part 2 of this manual is to provide students with a rudimentary knowledge of topographic maps, aerial photographs, and false color images so that they will be able to apply this knowledge in their study of landforms presented in Part 3.

Topographic Maps

Definition

A *topographic* map is a graphic representation of the three-dimensional configuration of the earth's surface. Most topographic maps also show land boundaries and other man-made features. The United States Geological Survey (U.S.G.S.), a unit of the Department of the Interior, has been actively engaged in the making of a series of standard topographic maps of the United States and its possessions since 1882.

Features of Topographic Maps

The features shown on topographic maps may be divided into three groups: (1) *relief,* which includes hills, valleys, mountains, plains, and the like; (2) *water features,* including lakes, ponds, rivers, canals, swamps, and the like; and (3) *culture,* works of man such as roads, railroads, land boundaries, and similar features (fig. 2.1). Relief is printed in brown; water in blue; and culture, including geographical names, in black; some recent maps now show major roads in red. (See fig. 2.2 for some standard map symbols.) On some maps, forests are printed in green.

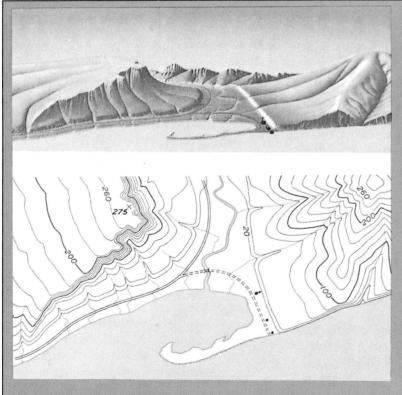

THE USE OF SYMBOLS IN MAPPING

These illustrations show how various features are depicted on a topographic map. The upper illustration is a perspective view of a river valley and the adjoining hills. The river flows into a bay that is partly enclosed by a hooked sandbar. On either side of the valley are terraces through which streams have cut gullies. The hill on the right has a smoothly eroded form and gradual slopes, whereas the one on the left rises abruptly in a sharp precipice from which it slopes gently, and forms an inclined table land traversed by a few shallow gullies. A road provides access to a church and two houses situated across the river from a highway that follows the seacoast and curves up the river valley.

The lower illustration shows the same features represented by symbols on a topographic map. The contour interval (the vertical distance between adjacent contours) is 20 feet.

Figure 2.1   Landforms as shown on a topographic map. Contour lines are printed in brown, water features in blue, and man-made structures in black. (Courtesy of the U.S.G.S.)

Figure 2.2   Standard symbols used on ⟶ topographic maps published by the U.S.G.S. (Courtesy of the U.S.G.S.)

Topographic Map Symbols

BOUNDARIES

National ....

State or territorial ....

County or equivalent ....

Civil township or equivalent ....

Incorporated-city or equivalent ....

Park, reservation, or monument ....

Small park ....

LAND SURVEY SYSTEMS

U.S. Public Land Survey System:

Township or range line ....

    Location doubtful ....

Section line ....

    Location doubtful ....

Found section corner; found closing corner ....

Witness corner; meander corner ....

Other land surveys:

Township or range line ....

Section line ....

Land grant or mining claim; monument ....

Fence line ....

ROADS AND RELATED FEATURES

Primary highway ....

Secondary highway ....

Light duty road ....

Unimproved road ....

Trail ....

Dual highway ....

Dual highway with median strip ....

Road under construction ....

Underpass; overpass ....

Bridge ....

Drawbridge ....

Tunnel ....

BUILDINGS AND RELATED FEATURES

Dwelling or place of employment: small; large ...

School; church ....

Barn, warehouse, etc.: small; large ....

House omission tint ....

Racetrack ....

Airport ....

Landing strip ....

Well (other than water); windmill ....

Water tank: small; large ....

Other tank: small; large ....

Covered reservoir ....

Gaging station ....

Landmark object ....

Campground; picnic area ....

Cemetery: small; large ....

RAILROADS AND RELATED FEATURES

Standard gauge single track; station ....

Standard gauge multiple track ....

Abandoned ....

Under construction ....

Narrow gauge single track ....

Narrow gauge multiple track ....

Railroad in street ....

Juxtaposition ....

Roundhouse and turntable ....

TRANSMISSION LINES AND PIPELINES

Power transmission line: pole; tower ....

Telephone or telegraph line ....

Aboveground oil or gas pipeline ....

Underground oil or gas pipeline ....

CONTOURS

Topographic:

Intermediate ....

Index ....

Supplementary ....

Depression ....

Cut; fill ....

Bathymetric:

Intermediate ....

Index ....

Primary ....

Index Primary ....

Supplementary ....

MINES AND CAVES

Quarry or open pit mine ....

Gravel, sand, clay, or borrow pit ....

Mine tunnel or cave entrance ....

Prospect; mine shaft ....

Mine dump ....

Tailings ....

SURFACE FEATURES

Levee ....

Sand or mud area, dunes, or shifting sand ......

Intricate surface area ....

Gravel beach or glacial moraine ....

Tailings pond ....

VEGETATION

Woods ....

Scrub ....

Orchard ....

Vineyard ....

Mangrove ....

COASTAL FEATURES

Foreshore flat ....

Rock or coral reef ....

Rock bare or awash ....

Group of rocks bare or awash ....

Exposed wreck ....

Depth curve; sounding ....

Breakwater, pier, jetty, or wharf ....

Seawall ....

BATHYMETRIC FEATURES

Area exposed at mean low tide; sounding datum ..

Channel ....

Offshore oil or gas: well; platform ....

Sunken rock ....

RIVERS, LAKES, AND CANALS

Intermittent stream ....

Intermittent river ....

Disappearing stream ....

Perennial stream ....

Perennial river ....

Small falls; small rapids ....

Large falls; large rapids ....

Masonry dam ....

Dam with lock ....

Dam carrying road ....

Intermittent lake or pond ....

Dry lake ....

Narrow wash ....

Wide wash ....

Canal, flume, or aqueduct with lock ....

Elevated aqueduct, flume, or conduit ....

Aqueduct tunnel ....

Water well; spring or seep ....

GLACIERS AND PERMANENT SNOWFIELDS

Contours and limits ....

Form lines ....

SUBMERGED AREAS AND BOGS

Marsh or swamp ....

Submerged marsh or swamp ....

Wooded marsh or swamp ....

Submerged wooded marsh or swamp ....

Rice field ....

Land subject to inundation ....

Map series and quadrangles

Each map in a U. S. Geological Survey series conforms to established specifications for size, scale, content, and symbolization. Except for maps which are formatted on a County or State basis, USGS quadrangle series maps cover areas bounded by parallels of latitude and meridians of longitude.

Map scale

Map scale is the relationship between distance on a map and the corresponding distance on the ground. Scale is expressed as a ratio, such as 1:25,000, and shown graphically by bar scales marked in feet and miles or in meters and kilometers.

Standard edition maps

Standard edition topographic maps are produced at 1:20,000 scale (Puerto Rico) and 1:24,000 or 1:25,000 scale (conterminous United States and Hawaii) in either 7.5 x 7.5- or 7.5 x 15-minute format. In Alaska, standard edition maps are available at 1:63,360 scale in 7.5 x 20 to 36-minute quadrangles. Generally, distances and elevations on 1:24,000-scale maps are given in conventional units: miles and feet, and on 1:25,000-scale maps in metric units: kilometers and meters.

The shape of the Earth's surface, portrayed by contours, is the distinctive characteristic of topographic maps. Contours are imaginary lines which follow the land surface or the ocean bottom at a constant elevation above or below sea level. The contour interval is the elevation difference between adjacent contour lines. The contour interval is chosen on the basis of the map scale and on the local relief. A small contour interval is used for flat areas; larger intervals are used for mountainous terrain. In very flat areas, the contour interval may not show sufficient surface detail and supplementary contours at less than the regular interval are used.

The use of color helps to distinguish kinds of features:

Black – cultural features such as roads and buildings.
Blue – hydrographic features such as lakes and rivers.
Brown – hypsographic features shown by contour lines.
Green – woodland cover, scrub, orchards, and vineyards.
Red – important roads and public land survey system.
Purple – features added from aerial photographs during map revision. The changes are not field checked.

Some quadrangles are mapped by a combination of orthophotographic images and map symbols. Orthophotographs are derived from aerial photographs by removing image displacements due to camera tilt and terrain relief variations. An orthophotoquad is a standard quadrangle format map on which an orthophotograph is combined with a grid, a few place names, and highway route numbers. An orthophotomap is a standard quadrangle format map on which a color enhanced orthophotograph is combined with the normal cartographic detail of a standard edition topographic map.

Provisional edition maps

Provisional edition maps are produced at 1:24,000 or 1:25,000 scale (1:63,360 for Alaskan 15-minute maps) in conventional or metric units and in either a 7.5 x 7.5- or 7.5 x 15-minute format. Map content generally is the same as for standard edition 1:24,000- or 1:25,000-scale quadrangle maps. However, modified symbolism and production procedures are used to speed up the completion of U.S. large-scale topographic map coverage.

The maps reflect a provisional rather than a finished appearance. For most map features and type, the original manuscripts which are prepared when the map is compiled from aerial photographs, including hand lettering, serve as the final copy for printing. Typeset lettering is applied only for features which are designated by an approved name. The number of names and descriptive labels shown on provisional maps is less than that shown on standard edition maps. For example, church, school, road, and railroad names are omitted.

Provisional edition maps are sold and distributed under the same procedures that apply to standard edition maps. At some future time, provisional maps will be updated and reissued as standard edition topographic maps.

National Mapping Program indexes

Indexes for each State, Puerto Rico, the U. S. Virgin Islands, Guam, American Samoa, and Antarctica are available. Separate indexes are available for 1:100,000-scale quadrangle and county maps; USGS/Defense Mapping Agency 15-minute (1:50,000-scale) maps; U. S. small scale maps (1:250,000, 1:1,000,000, 1:2,000,000 scale; State base maps; and U. S. maps); land use/land cover products; and digital cartographic products.

Series	Scale	1 inch represents approximately	1 centimeter represents	Size (latitude x longitude)	Area (square miles)
Puerto Rico 7.5-minute	1:20,000	1,667 feet	200 meters	7.5 x 7.5 min.	71
7.5-minute	1:24,000	2,000 feet (exact)	240 meters	7.5 x 7.5 min.	49 to 70
7.5-minute	1:25,000	2,083 feet	250 meters	7.5 x 7.5 min.	49 to 70
7.5 x 15-minute	1:25,000	2,083 feet	250 meters	7.5 x 15 min.	98 to 140
USGS/DMA 15-minute	1:50,000	4,166 feet	500 meters	15 x 15 min.	197 to 282
15-minute	1:62,500	1 mile	625 meters	15 x 15 min.	197 to 282
Alaska 1:63,360	1:63,360	1 mile (exact)	633.6 meters	15 x 20 to 36 min.	207 to 281
County 1:50,000	1:50,000	4,166 feet	500 meters	County area	Varies
County 1:100,000	1:100,000	1.6 miles	1 kilometer	County area	Varies
30 x 60-minute	1:100,000	1.6 miles	1 kilometer	30 x 60 min.	1,568 to 2,240
U. S. 1:250,000	1:250,000	4 miles	2.5 kilometers	1° x 2° or 3°	4,580 to 8,669
State maps	1:500,000	8 miles	5 kilometers	State area	Varies
U. S. 1:1,000,000	1:1,000,000	16 miles	10 kilometers	4° x 6°	73,734 to 102,759
U. S. Sectional	1:2,000,000	32 miles	20 kilometers	State groups	Varies
Antarctica 1:250,000	1:250,000	4 miles	2.5 kilometers	1° x 3° to 15°	4,089 to 8,336
Antarctica 1:500,000	1:500,000	8 miles	5 kilometers	2° x 7.5°	28,174 to 30,462

How to order maps

Mail orders. Order by map name, State, and series/scale. Payment by money order or check payable to the U. S. Geological Survey must accompany your order. Your complete address, including ZIP code, is required.

Maps of areas *east* of the Mississippi River, including Minnesota, Puerto Rico, the Virgin Islands of the United States, and Antarctica.

Eastern Distribution Branch
U. S. Geological Survey
1200 South Eads Street
Arlington, VA 22202

Maps of areas *west* of the Mississippi River, including Alaska, Hawaii, Louisiana, American Samoa, and Guam.

Western Distribution Branch
U. S. Geological Survey
Box 25286, Federal Center
Denver, CO 80225

Standard topographic maps of the U.S.G.S. cover a *quadrangle* of area that is bounded by *parallels of latitude* (forming the northern and southern margins of the map) and by *meridians of longitude* (forming the eastern and western margins of the map). The published maps have different *scales*. A map scale is a means of showing the relationship between the size of an object or feature indicated on a map and the corresponding actual size of the same object or feature on the ground.

Geologists make use of topographic maps because they provide them with a means to observe earth features in *three dimensions*. Unlike other maps, topographic maps show natural features to a fair degree of accuracy in terms of length, width, and vertical height or depth. Thus by examination of a topographic map and through an understanding of the symbols shown thereon, geologists are able to interpret earth features and draw conclusions as to their origin in the light of geologic processes.

In the United States, Canada, and other English-speaking countries of the world, most maps produced to date have used the English system of measurement. That is to say, distances are measured in feet, yards, or miles; elevations are shown in feet; and water depths are recorded in feet or fathoms (1 fathom = 6 feet). In 1977, in accordance with national policy, the U.S.G.S. formally announced its intent to convert all of its maps to the metric system. As resources and circumstances permit, new maps published by the U.S.G.S. will show distances in kilometers and elevations in meters. The conversion from English to metric units will take many decades in the United States. In this manual, most maps used will be those published by the U.S.G.S. *prior* to the adoption of the metric system, because the metric maps are still insufficient in number to portray the great diversity of geologic features presented in this manual. However, some examples of map scales in metric units will be given to acquaint students with the system.

To help students familiarize themselves with the metric system and its relationship to the English system, table 2.1 is provided as a convenient reference.

Elements of a Topographic Map

Map Scale

Three scales are commonly used in conjunction with topographic maps: (1) *fractional,* (2) *graphic,* and (3) *verbal.*

1. A *fractional scale* is a fixed ratio between linear measurements on the map and corresponding distances on the ground. It is sometimes called the *representative fraction* or R. F.

$$\text{Example: R. F. 1:62,500 or } \frac{1}{62,500}.$$

Table 2.1. Units of Measurement in the English and Metric Systems, and the Means of Converting from One to the Other

A. English Units of Linear Measurement
    12 inches = 1 foot
    3 feet = 1 yard
    1 mile = 1,760 yards, 5,280 feet, 63,360 inches

B. Metric Units of Linear Measurement
    10 millimeters = 1 centimeter
    100 centimeters = 1 meter
    1,000 meters = 1 kilometer

C. Conversion of English Units to Metric Units

symbol	when you know	multiply by	to find	symbol
in.	inches	2.54	centimeters	cm
ft.	feet	30.48	centimeters	cm
yd.	yards	0.91	meters	m
mi.	miles	1.61	kilometers	km

D. Conversion of Metric Units to English Units

symbol	when you know	multiply by	to find	symbol
mm	millimeters	0.04	inches	in.
cm	centimeters	0.4	inches	in.
m	meters	3.28	feet	ft.
m	meters	1.09	yards	yd.
km	kilometers	0.62	miles	mi.

This notation simply means that 1 unit on the map equals 62,500 of the *same units* on the ground. Thus, a line 1 inch long on the map represents a horizontal distance of 62,500 inches on the ground. (Note that the numerator of the R. F. is always 1.)

2. A *graphic scale* is simply a line or bar drawn on the map and divided into units that represent ground distances.

3. A *verbal scale* is a convenient way of stating the relationship of map distance to ground distance. For example, "1 inch equals 1 mile" is a verbal scale and means that 1 inch on the map equals 1 mile on the ground. Or, "1 cm equals 1 km" means that 1 centimeter on the map equals 1 kilometer on the ground.

Figure 2.3 provides a visual comparison of three maps of Mt. Rainier, Washington, at scales of 1:24,000, 1:100,000, and 1:125,000.

Converting from One Scale to Another

It is sometimes necessary to convert from a fractional scale to a verbal or graphic scale, or from verbal to fractional. This involves simple problems in arithmetic. The following are examples of some conversion problems.

Example 1. Convert an R. F. of 1:125,000 to a verbal scale in terms of inches per mile. In other words, how many miles on the ground are represented by 1 inch on the map?

Solution: It is best to express the R. F. as an equation in terms of what it actually means. Thus,

1 unit on the map = 125,000 units on the ground.

1:24,000-scale

A

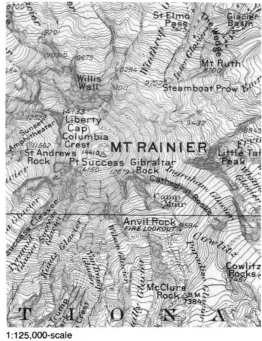

1:125,000-scale

C

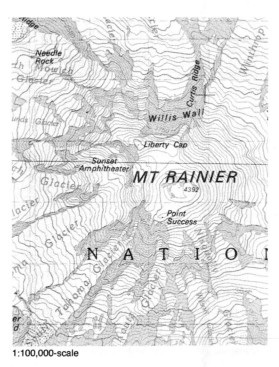

1:100,000-scale

B

Figure 2.3 Maps of Mt. Rainier at various scales. (A) 1:24,000 scale, (B) 1:100,000 scale, (C) 1:125,000 scale. (Courtesy U.S.G.S. National Mapping Program Pamphlet, *Map Scales*, 1981.)

We see from this basic equation that we are dealing in the same units on the map and on the ground. Since we are interested in *1 inch* on the map, let us substitute *inches* for *units* in the above equation. It then reads:

1 inch on the map = 125,000 inches on the ground.

The left side of the equation is now complete, since we initially wanted to know the ground distance represented by *1 inch* on the map. We have yet to resolve the right side of the equation into miles because the problem specifically called for a ground distance in miles. Since there are 5,280 feet in a mile and 12 inches in a foot,

$$5,280 \times 12 = 63,360 \text{ inches in one mile.}$$

If we divide 125,000 inches by the number of inches in a mile, we arrive at:

$$\frac{125,000}{63,360} = 1.97 \text{ miles.}$$

The answer to our problem is one inch = 1.97 miles or, for all practical purposes, 2 miles.

Example 2. On a certain map, 2.8 inches is equal to 1.06 miles on the ground. Express this verbal scale as an R. F.

Solution: Again, it is wise to follow simple steps to avoid confusion. We know that, by definition,

$$\text{R. F.} = \frac{\text{distance on the map}}{\text{distance on the ground}}.$$

Since we were given a map distance of 2.8 inches and the equivalent ground distance of 1.06 miles, it is a simple matter to substitute these in our basic equation. Thus,

$$R. F. = \frac{2.8 \text{ inches on the map}}{1.06 \text{ miles on the ground}}.$$

It is now necessary to change the denominator of the equation to inches because both terms in the R. F. must be expressed in the same units. Since there are 63,360 inches in 1 mile, we must multiply the denominator by that number:

$$1.06 \times 63,360 = 67,161.6 \text{ inches}.$$

Now our equation reads,

$$R. F. = \frac{2.8 \text{ inches}}{67,161.6 \text{ inches}}.$$

To complete the problem, we need to express the numerator as unity, and so the numerator must be divided by itself. In order not to change the value of the fraction, we must also divide the denominator by the same number, 2.8. The problem resolves itself into,

$$R. F. = \frac{2.8 \div 2.8}{67,161.6 \div 2.8} = \frac{1}{23,986}$$

or rounded off to $\frac{1}{24,000}$.

If you can follow these two examples and understand them completely, you can handle any type of conversion problem.

Exercise 7. Problems in Scale Conversion

1. Convert the following representative fractions to ground distances equal to one inch on a map. (Fill in the following blanks.)
   a) 1:24,000   1 inch = _____ feet
   b) 1:31,680   1 inch = _____ miles
   c) 1:48,000   1 inch = _____ miles
   d) 1:62,500   1 inch = _____ miles
   e) 1:250,000   1 inch = _____ miles
   f) 1:1,000,000   1 inch = _____ miles
   g) 1:1,000,000   1 cm   = _____ km

2. A map of unknown scale shows two TV transmitting towers. On the map the towers are 1.2 inches apart and the actual ground distance between them is 1,000 feet. What is the R. F. of the map?

3. A straight stretch of road on an aerial photograph was found to be 500 yards long. The same road segment measured on the photograph was three-fourths of an inch. What is the R. F. of the photograph?

4. A foreign map has an R. F. of 1:500,000. How many kilometers on the ground are represented by 10 centimeters on the map?

5. The bar scales shown in figure 2.4 appear on one of the new U.S.G.S. maps. Derive the R. F. of the map first using the kilometer scale and then using the mile scale. Which of the two systems, English or metric, is simpler to handle in the derivation of the R. F.? Why?

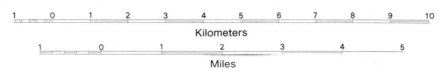

Figure 2.4   Two bar scales from a recent U.S.G.S. topographic map.

Map Coordinates and Land Divisions

The earth's surface is arbitrarily divided into a system of reference coordinates called *latitude* and *longitude*. This coordinate system consists of imaginary lines on the earth's surface called *parallels* and *meridians* (fig. 2.5). Both of these are best described by assuming the earth to be represented by a globe with an axis of rotation passing through the North and South poles. Meridians are circles drawn on this globe that pass through the two poles. Meridians are labeled according to their positions, in degrees, from the zero meridian, which by international agreement passes through Greenwich near London, England. The zero meridian is commonly referred to as the *Greenwich meridian*. If meridional lines are drawn for each 15 degrees in an easterly direction from Greenwich (toward Asia) and in a westerly direction from Greenwich (toward North America), a family of great circles will be created. Each one of the great circles is labeled according to the number of degrees it lies east or west of the Greenwich or zero meridian. The 180° west meridian and the 180° east meridian are one and the same great circle, and constitute the International Date Line.

A great circle represents the intersection of a plane that connects two points on the surface of a sphere and passes through the center of the sphere, in this case the earth. The intersection of the plane with the surface divides the earth into two equal halves—hemispheres—and the arc of the great circle is the shortest distance between two points on the spherical earth.

Another great circle passing around the earth midway between the two poles is the equator. It divides the earth into the Northern and Southern Hemispheres. A family of lines drawn on the globe parallel to the equator constitute the second set of reference lines needed to locate a point on the earth accurately. These lines form circles that are called *parallels of latitude,* labeled according to their distances in degrees north or south of the equator. The parallel that lies halfway between the equator and the North Pole is latitude 45° North, and the North Pole itself lies at latitude 90° North.

This system of meridians and parallels thus provides a means of accurately designating the location of any point on the globe. Santa Monica, California, for example, lies at about longitude 118° 29' West and latitude 34° 01' North. For increased accuracy in locating a point, degrees may be subdivided into 60 subdivisions known as *minutes,* indicated by the notation '. Minutes may be subdivided into 60 subdivisions known as *seconds,* indicated by the notation ''. Thus, a position description might read 64° 32' 32'' East, 44° 16' 18'' South.

Meridional lines converge toward the North or South Pole from the equator, and the length of a degree of longitude varies from 69.17 statute miles at the equator to zero at the poles. Latitudinal lines, on the other hand, are always parallel to each other. However, because the earth

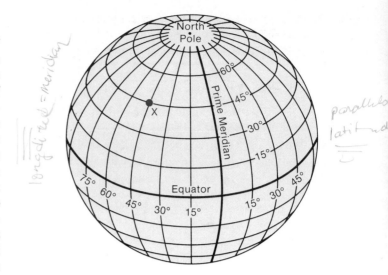

Figure 2.5 Generalized system of meridians and parallels. The location of point X is longitude 45° West, latitude 45° North.

is not a perfect sphere but is slightly bulged at the equator, a degree of latitude varies from 68.7 statute miles at the equator to 69.4 statute miles at the poles. Thus, the area bounded by parallels and meridians is not a true rectangle. U.S.G.S. quadrangle maps are also bounded by meridians and parallels, but on the scale at which they are drawn, the convergence of the meridional lines is so slight that the maps appear to be true rectangles. The U.S.G.S. standard quadrangle maps embrace an area bounded by 7½ minutes of longitude and 7½ minutes of latitude. These quadrangle maps are called 7½-minute quadrangles. Other maps published by the U.S.G.S. are 15-minute quadrangles, and a few of the older ones are 30-minute quadrangles.

Meridians always lie in a true north-south direction, and parallels always lie in a true east-west direction. *Magnetic north,* however, is the direction toward which the north-seeking end of a magnetic compass needle will point. Because the magnetic poles are not coincident with the north and south ends of the earth's rotational axis, magnetic north is different from true north except on the meridian that passes through the magnetic North Pole. The angle between true north and magnetic north is called the *declination,* and is normally shown on the lower margin of most U.S.G.S. maps for the benefit of those who use a compass in the field to plot geological or other data on a base map (e.g., a U.S.G.S. standard quadrangle map).

Map Projection

While it is easy to draw the system of meridians and parallels on a globe, it is not possible to draw this system on a flat piece of paper (a map) without introducing some distortion. A wide variety of methods have been developed to reduce this distortion during the construction on a map.

This process of constructing a map, the transferring of the meridians and parallels to a flat sheet of paper, is a geometric exercise called *projection,* and the resulting product is called a *map projection.* The map projection selected for use by the cartographer, one who draws maps, depends on the purpose of the map and the material to be presented.

The variety of map projections available include a group of projections that preserve the property of area, *equal-area* projections. On an equal-area a coin placed any place of the map will cover the same area. These projections distort shape and the angular relations between meridians and parallels.

Conformable maps represent a group of projections that preserve shape but greatly distort area. Because many publications prefer to present maps that preserve shape, readers often have a mistaken impression of the relative size of the continents. An example of this is the size of Greenland, actually about the same size as Mexico, but on a conformable map of the world it appears many times larger. There are a variety of projections that preserve angular relationships between the meridians and parallels or that, while preserving none of these properties, are compromises used for specific purposes. Most of the U.S.G.S. maps found in this manual are drawn on a polyconic projection. This projection preserves neither shape nor area, but for the small area represented the distortion of both is at a minimum. The 1:250,000 scale map of Greenville (Fig. 3.10) uses a transverse Mercator projection, and figure 4.36 uses a Lambert conformal conic projection.

The map used for figure 5.1 is a Mercator projection, a commonly used conformable projection that has the advantage of preserving shape and having the same scale in all directions near any given point, along the equator, or along two parallels equidistant from the equator. The disadvantage of this projection is the great distortion of area in the polar regions and the fact that the very high latitudes and poles themselves cannot be shown.

Range and Township

For purposes of locating property lines and land descriptions on legal documents, another system of coordinates is used in the United States and some parts of Canada. This system is tied into the latitude and longitude coordinate system, but functions independently of it. The basic block of this system is the *section,* a rectangular block of land one mile long and one mile wide. An area containing 36 sections is called a *township.* While it was the intent of the original government land surveyors to make each section an exact square of land, many sections and townships are irregular in shape because of surveying errors and other discrepancies in laying out the network.

The north-south lines marking township boundaries are called *range lines,* and the east-west boundaries are

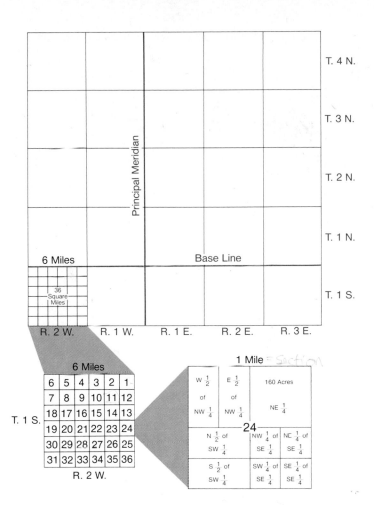

Figure 2.6   Standard land divisions used in the United States and some parts of Canada.

called *township lines.* The coordinate system of numbering townships has as a reference or beginning point, the intersection of a meridian of longitude and a parallel of latitude, called *principal meridian* and *baseline,* respectively. A particular township is identified by stating its position north or south of the baseline and east or west of the principal meridian. The system of numbering township and range lines is shown in figure 2.6. The letter *T* along the right-hand margin of the large map stands for the word *township,* and the letter *R* stands for the word *range.* The notation T. 3 N., R. 1 E. is read, "Township three north, Range one east." Under this system, each township has a unique numerical designation. Several principal meridians and baselines are used in the coterminous United States, so that the township and range coordinate numbers are never very large.

Each township consists of 36 one-square-mile sections that are numbered according to the system shown in figure 2.6. One section of land contains 640 acres.

For purposes of locating either man-made or natural features in a given section, an additional convention is employed. This consists of dividing the section into quarters called the northeast quarter (NE¼), the southwest quarter (SW¼), and so on. Sections may also be divided into halves such as the north half (N½) or west half (W½). The quarter sections are further divided into four more quarters or two halves, depending on how refined one wants to make the description of a feature on the ground. For example, an exact description of the 40 acres of land in the extreme southeast corner of the map of section 24 in figure 2.6 would be as follows: SE¼ of the SE¼ of Section 24, T. 1 S., R. 2 W. This style of notation will be used throughout the manual in referring the student to a particular feature on a map used in an exercise.

Figure 2.7 is an aerial photograph of rural Iowa. The rectangular network of roads tends to follow section lines, and the cultivated fields generally conform in shape and size according to the system of land divisions shown in figure 2.6.

Because the maps used in this manual are only *selected parts* of standard U.S.G.S. quadrangle maps, the township and range lines numbering system, magnetic declination, and other data usually found on the lower margin of the maps may not be included. Where necessary, these data will be supplied in the text of the exercise.

Topography

Topography is the configuration of the land surface and is shown by means of *contour lines* (fig. 2.1). A contour line is an imaginary line on the surface of the earth connecting points of equal elevation. The *contour interval* (C.I.) is the difference in elevation of any two adjacent contour lines. Elevations are given in feet or meters above mean sea level. The shore of a lake is, in effect, a contour line because every point on it is at the same level (elevation).

Contour lines are brown on the standard U.S.G.S. maps. The C.I. is usually constant for a given map and may range from 5 feet for flat terrain to 50 or 100 feet for a mountainous region. C.I.'s for U.S.G.S. maps using the metric system are 1, 2, 5, 10, 20, 50, or 100 meters, depending upon the smoothness or ruggedness of the terrain

to be depicted. Usually, every fifth contour line is printed in a heavier line than the others and bears the elevation of the contour above sea level. In addition to contour lines, the heights of many points on the map, such as road intersections, summits of hills, and lake levels are shown to the nearest foot or meter on the map. These are called *spot elevations* and are accurate to within the nearest foot or meter. More precisely located and more accurate in elevation are *bench marks,* points marked by brass plates permanently fixed on the ground, and by crosses and elevations, preceded by the letters *BM,* printed in black on the map.

Following are several general statements regarding contour lines.

1. When contour lines cross streams, they bend upstream; that is, the segment of the contour line near the stream forms a "V" with the apex pointing in an upstream direction.
2. Contours do not intersect or cross unless they become merged in a vertical or overhanging cliff.
3. Closed contours appearing on the map as ellipses or circles represent hills or knobs.
4. Closed contours with hachures, short lines pointing toward the center of the closure (downslope), represent closed depressions. The outer hachured contour line has the same elevation as the lower adjacent regular contour line.
5. Steep slopes are shown by closely spaced contours, gentle slopes by widely spaced contours.
6. The difference in elevation between the highest and the lowest point of a given area is the *maximum relief* of that area.

Examine figure 2.1 for the relationship of contour lines to topography. In particular, note the difference in spacing between the contour lines that depict the steep slope on the left side of the river and the contour lines that depict the gentle slope on the right side of the river. Where the cliff is almost vertical in the lower left-hand corner of the map, the contour lines are almost "stacked" one on top of the other.

Figure 2.7 Aerial photograph of a rural area in Iowa. The town is Storm Lake. (Courtesy of the U.S.G.S. Photo was taken on September 23, 1950.)

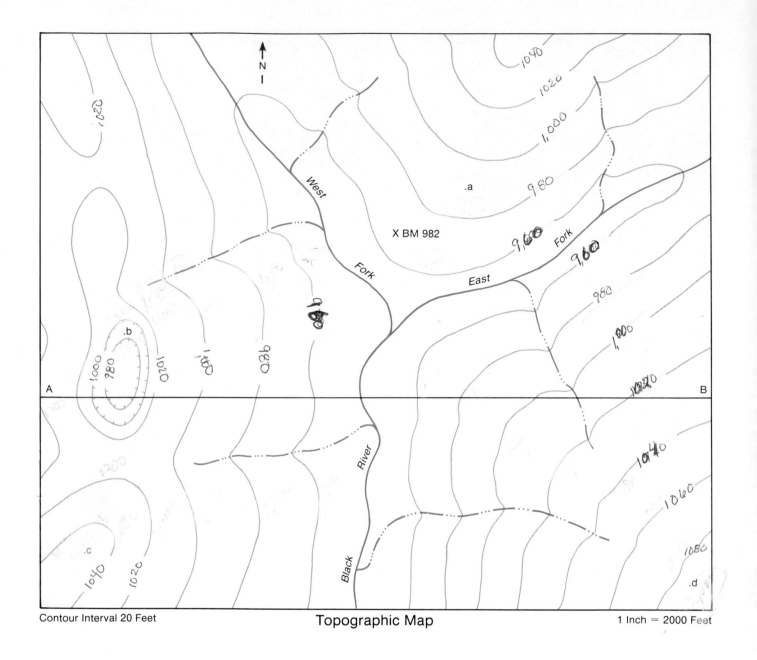

Contour Interval 20 Feet   Topographic Map   1 Inch = 2000 Feet

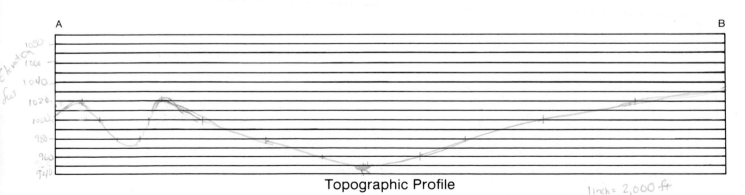

Topographic Profile

RF = 1:24.000   1 inch = 2,000 ft

Figure 2.8   Topographic map and topographic profile to be used in connection with Exercise 8, question 1, and Exercise 9, question 1.

Exercise 8. Problems on Contour Lines

1. Study the map in the upper part of figure 2.8. Note the elevation of the bench mark (B.M. 982). This notation means that the point marked X is 982 feet above sea level. Label each contour line with its proper elevation and determine the approximate elevation of points labeled *a, b, c,* and *d.* (Note that the C.I. is 20 feet.) The elevation of each contour line should be written in the space between the broken contour lines (see fig. 2.1 for example).

2. The map shown in figure 2.9 shows spot elevations, drainage lines, and a lake. Using a C.I. of 5 feet, construct a topographic map of the area. (Note: All of the spot elevations will not lie on one of the lines you draw. To draw the contour lines, you will have to interpolate where the lines should be drawn between points of known elevation. This requires some judgment as to the placement of the lines. For example, only one spot elevation on the map is shown at the 65-foot elevation. To draw the 65-foot contour line you must estimate where other points at that elevation lie relative to known points of elevation. Thus, the 65-foot contour would probably be drawn closer to a spot elevation of 64 feet than it would to a spot elevation of 70 feet.)

3. What is the R. F. of the map in figure 2.8?

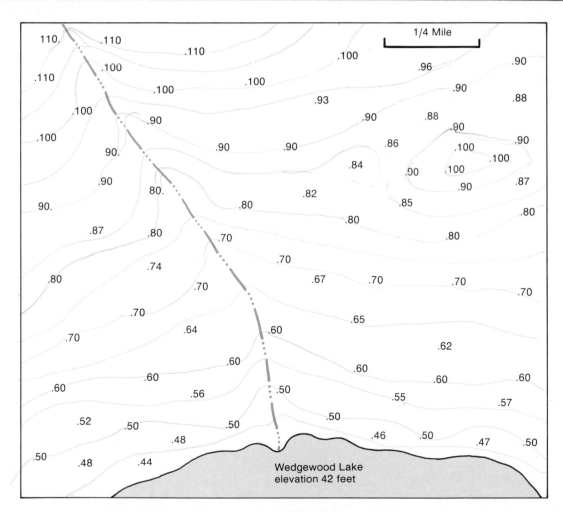

Figure 2.9 Map showing drainage lines and spot elevations to be used with Exercise 8, question 2.

Topographic Profiles

A *topographic profile* is a diagram that shows the change in elevation of the land surface along any given line. It represents graphically the "skyline" as viewed from a distance. Features shown in profile are viewed along a horizontal line of sight, whereas features shown on a map or in *plan view* are viewed along a vertical line of sight. Topographic profiles can be constructed from a topographic map along any given line.

The *vertical scale* of a profile is arbitrarily selected and is usually, but not always, larger than the horizontal scale of the map from which the profile is drawn. Only when the horizontal and vertical scales are the same is the profile *true,* but in order to facilitate the drawing of the profile and to emphasize differences in relief, a larger vertical scale is used. Such profiles are *exaggerated profiles.*

Ideally both the horizontal and vertical scales would be the same. This is impractical in most cases. On a section with a horizontal scale of 1:1,000,000, the topographic features would be almost impossible to see if the same vertical scale were used. Both horizontal and vertical scales are usually provided with each profile. The exaggeration is determined by comparing the inches on the profile with feet in nature. Thus, on a profile with a horizontal scale of 1:24,000 and a vertical scale of 1/10 inch to every 100 feet of elevation:

Horizontally 1 inch represents 24,000 divided by 12 = 2,000 ft.
Vertically 1 inch represents 10 times 100 = 1,000 ft.
Vertical exaggeration is 2.0 times.

Be aware that vertical exaggeration not only increases but also changes the character of the profile. A volcano such as that shown in figure 2.14 would appear as a sharp peak at 10 times vertical exaggeration.

Instructions for Drawing a Topographic Profile

Figure 2.10 shows the relationship of a topographic profile to a topographic map. It should be examined in connection with the following instructions:

1. The line along which a cross section is to be constructed may be defined by an actual line drawn on the map, or by two points on the map that determine the terminals of the line of cross section.

2. Examine the line along which the profile is to be drawn and note the difference between the highest and lowest contours crossed by it. The difference between them is the maximum relief of the profile. Cross sectional paper divided into 0.1 inch or 2 mm squares makes a good base on which to draw a profile. Use a vertical scale as small as possible so as to keep the amount of vertical exaggeration to a minimum. For example, for a profile along which the maximum relief is less than 100 feet, a vertical scale of 0.1 inch or 2 mm = 5, 10, 20, or 25 feet is appropriate. For a profile with 100 to 500 feet of maximum relief, a vertical scale of 0.1 inch or 2 mm = 40 or 50 feet is proper. If the maximum relief along the profile is between 500 and 1,000 feet, a vertical scale of 0.1 inch or 2 mm = 80 or 100 feet is adequate. When the maximum relief is greater than 1,000 feet, a vertical scale of 0.1 inch or 2 mm = 200 feet is appropriate. The general rule for guidance in the selection of a vertical scale is: the greater the maximum relief, the smaller the scale. Label the horizontal lines of the profile grid with appropriate elevations from the contours crossed by the line of the profile. Every other line on a 0.1 inch or 2 mm grid is sufficient.

3. Place the edge of the cross sectional paper along the line of profile. Opposite each intersection of a contour line with the line of profile, mark a short dash at the edge of the cross sectional paper. If the contour lines are closely spaced, only the heavy or *index* contours need to be marked. Also mark the positions of streams, lakes, hilltops, and significant cultural features on the line of profile. At the edge of the paper, label the elevation of each dash.

4. Drop these elevations perpendicularly to the corresponding elevations represented by the horizontal lines on the cross sectional paper.

5. Connect these points by a smooth line and label significant features such as streams and summits of hills. Add the horizontal scale and write a title on the profile.

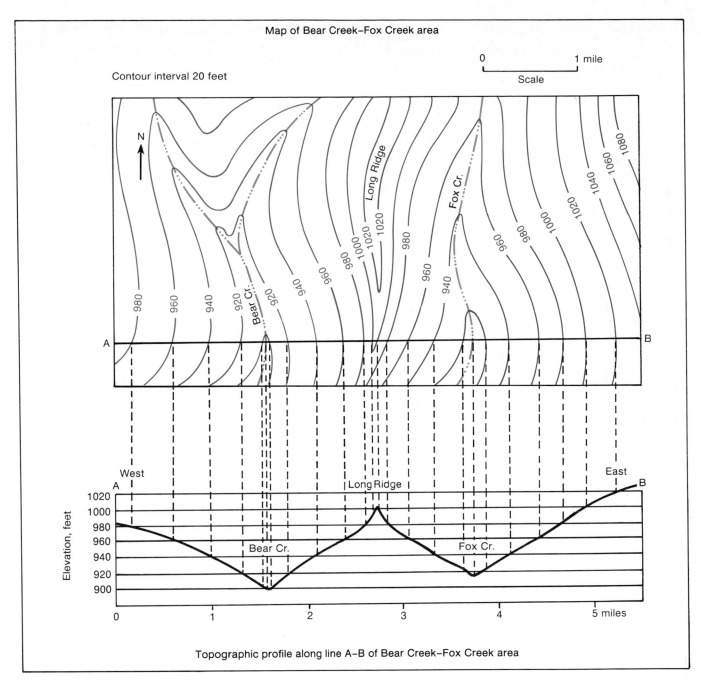

Figure 2.10 A topographic profile drawn along line A–B on the map of the hypothetical Bear Creek—Fox Creek area. See text for step-by-step instructions.

k2 copy for grid

Exercise 9. Drawing Profiles from Topographic Maps

1. Draw a profile along line A–B of the topographic map in figure 2.8.

2. Refer to the Delaware Map of figure 2.12. Three north-south red lines and three east-west red lines intersect at approximately one-mile intervals. These north-south and east-west lines define the boundaries of *sections*. Each section is numbered, and the number is printed in red in the center of each section.

   On the grid of figure 2.11, draw a north-south profile (in pencil) along the red line that defines the western boundary of sections 10, 3, and 34. The beginning point of the profile is the southwest corner of section 10, and the ending point is the shore of Lake Superior, which has an elevation of 602 feet above sea level. (Part of Lake Superior is indicated along the northern part of the Delaware Map.) The vertical scale has been established as

0.1 inch = 40 feet. The horizontal scale is the same as the map scale. Significant features along the line of profile have already been labeled.

3. What is the horizontal scale of the profile in feet per inch?

4. What is the vertical scale of the profile you have drawn in feet per inch?

5. By what factor is the vertical scale exaggerated?

6. In order to visualize more clearly the effect of vertical exaggeration, redraw the profile using a vertical scale of 0.1 inch = 80 feet. Use the grid in figure 2.11 and label the horizontal lines (according to the new scale) on the right-hand margin of the grid. The horizontal line now labeled 600 feet will be labeled 800, and the line now labeled 800 will be 1,200 on the new scale.

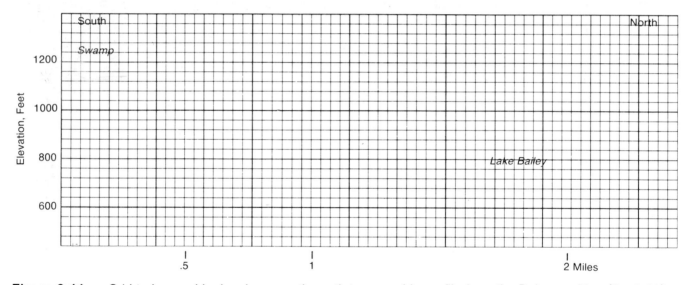

Figure 2.11   Grid to be used in drawing a north-south topographic profile from the Delaware Map (fig. 2.12). The south end of the profile is the southwest corner of section 10, and the north end is the shore of Lake Superior. The line of profile is coincident with the western boundaries of sections 3, 10, and 24.

Exercise 10. Topographic Map Reading

The following questions are based on the Delaware Map, figure 2.12.

1. What is the R. F. of the map?

2. What is the C.I. of the map?

3. If the C.I. was twice what it actually is, would there be a greater or lesser number of contour lines?

4. What is the map distance in feet between the highest point on Mt. Lookout (1,335 feet) and the nearest point on the south shore of Lake Bailey?

5. If you walked along the line described in question 4, would the actual distance walked be greater, less, or the same as the map distance? Explain.

6. What is the maximum relief of the map area?

7. Give the location of the following features using the township, range, section, etc. (the northern-most east-west red line separates T. 58 N. from T. 59 N., and the entire map area falls in R. 30 W.).
   a) North Pond.
   b) Confluence of Bailey Creek with Silver River.

8. Refer to figure 2.6. Notice that each square mile of land (1 section) is comprised of 640 acres. Each quarter-section contains 160 acres ($\frac{1}{4} \times 640 = 160$). Knowing that 160 acres = one quarter-section and that a quarter-section of land is a tract $\frac{1}{2}$ mile (2,640 feet) on a side, determine
   a) the number of square feet in a quarter-section (160 acres).
   b) the number of square feet in 1 acre.

9. Approximately how many acres are covered by Lake Bailey? (*Suggestion:* First determine the area of the lake in square feet by multiplying the length of the long axis by the length of the short axis. Do the same for the island. Subtract the area of the island from the area of the lake and convert the difference to acres.)

10. What is the direction of flow of the intermittent stream located in sections 10 and 11?

11. What is the elevation of the contour line that surrounds the small pond between Silver River and Highway 26 just south of the shore of Lake Superior? (Lake Superior is the large body of water across the northern part of the map.)

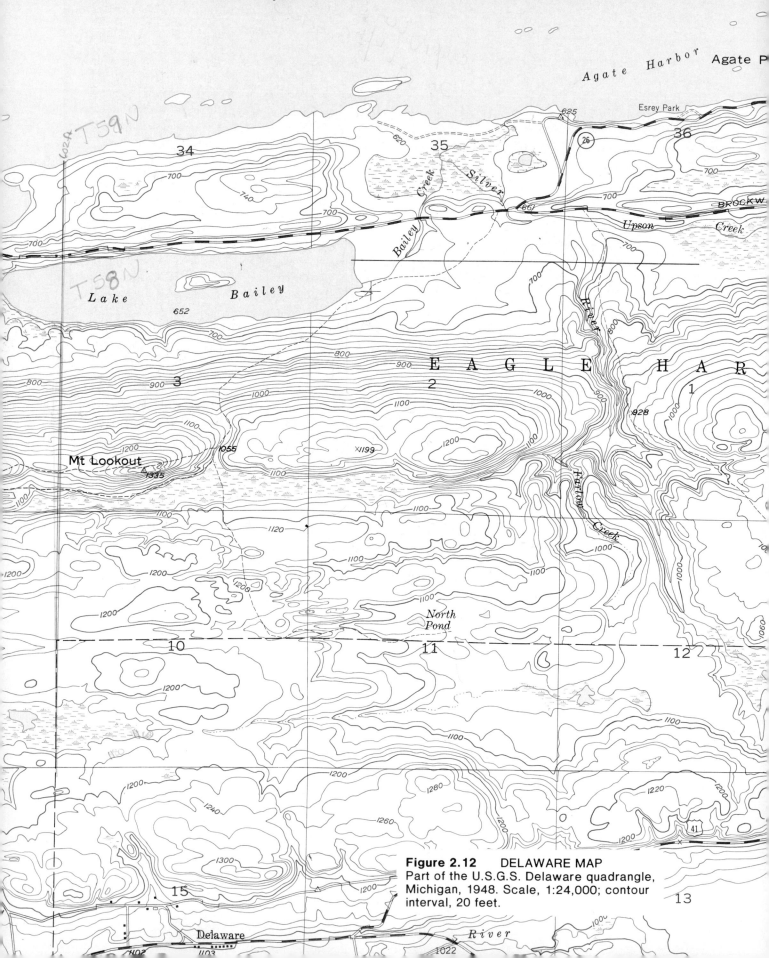

Figure 2.12   DELAWARE MAP
Part of the U.S.G.S. Delaware quadrangle,
Michigan, 1948. Scale, 1:24,000; contour
interval, 20 feet.

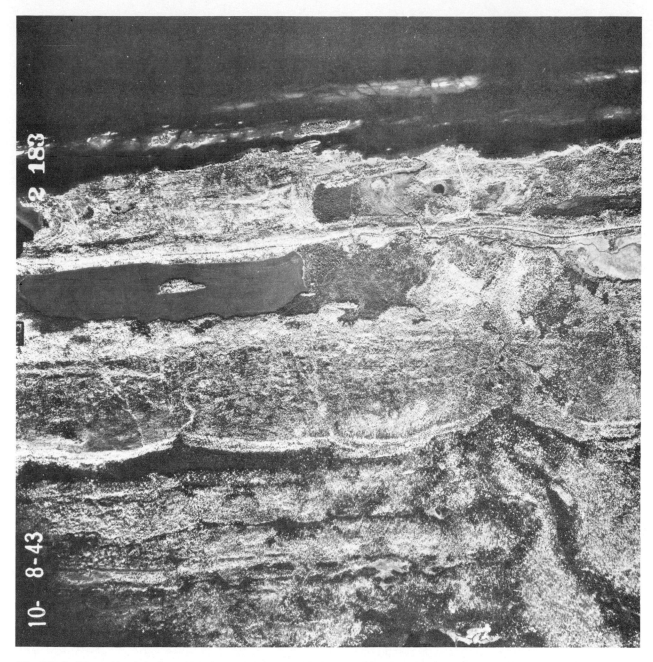

Figure 2.13   Aerial photograph of part of the Delaware Map on the opposite page. The photo is shown here for comparative purposes and will be used in a later exercise. (Courtesy of the U.S.G.S. Photo was taken on October 8, 1943.)

Aerial Photographs and Other Imagery from Remote Sensing

Background

An aerial photograph is a picture taken from an airplane flying at altitudes as high as 60,000 feet. Practically all of the United States has been photographed aerially by federal, state, or other agencies.

Photographs taken with the axis of the camera pointed vertically down are *vertical photos* and are most useful to the geologist. Air photos are also used for the making of topographic maps, for highway location, military operations, mapping of soils, city planning, and for many other operations.

Other "pictures," made with special sensors mounted in earth-orbiting satellites, are also useful for displaying features of the earth's surface. The false color images are perhaps the most spectacular from an aesthetic point of view, but they are quite useful in geologic interpretative work because of the broad regional relationships they show. False color images and other kinds of earth satellite imagery have been made with the imaging devices aboard the several Landsat satellites, the first of which was launched on July 23, 1973. Imagery of a variety of types including side looking radar has been acquired from devices on board other satellites or space vehicles. The satellite data is transmitted to receiving stations on earth. These data are then processed into images of various types and are made available to users through commercial sources or agencies of the federal government such as the U.S.G.S. and NOAA.

Space imagery results in "maps" that are very small in scale; that is, an inch on these image maps represents many miles on the ground. Thus, the Landsat maps do not have the resolution (degree of detail) that aerial photographs have. For this reason, most of the remote sensing imagery used in this manual will consist of aerial photographs. Where appropriate, a false color image will be used.

Determining Scale on Aerial Photographs

The scale of an air photo depends on both the focal length of the camera and the height of the airplane above ground surface. If these two factors are known, the R. F. of the photo can be determined. The photo scale can also be determined, however, by measuring known ground distances on the photo. In many parts of the coterminous United States, section lines are ideally suited for this purpose since they are normally 1 mile apart and are usually visible on the photo as roads, fences, or other man-made features. Conversion to an R. F. or graphic scale can be made in the same manner as was described under the section on topographic map scales.

Stereoscopic Use of Aerial Photographs

If two vertical photos are taken from a slightly different position and viewed through a *stereoscope,* the relief of the land becomes visible. A *stereopair* consists of two photos viewed in such a way that each of the eyes sees only one of the two photos. The brain combines the two images to form a three-dimensional view of the objects shown on the photograph.

Figure 2.14 is a stereopair for practice in using the simple lens stereoscope. Figure 2.15 shows one being used. The stereoscope is positioned over the stereopair so that the viewer's nose is directly over the line separating the two adjacent photos. If stereovision is not immediately achieved, the stereoscope should be rotated slightly around an imaginary vertical axis passing through the midpoint between the two lenses until the image viewed appears in relief.

On photographs in which steep slopes occur (e.g., the walls of deep canyons or the flanks of high mountains), the stereoscopic image of these features is exaggerated. That is, steeply inclined canyon walls may appear to be nearly vertical when, in fact, they are not. This distortion, however, does not present a problem under normal conditions of viewing by the beginning student.

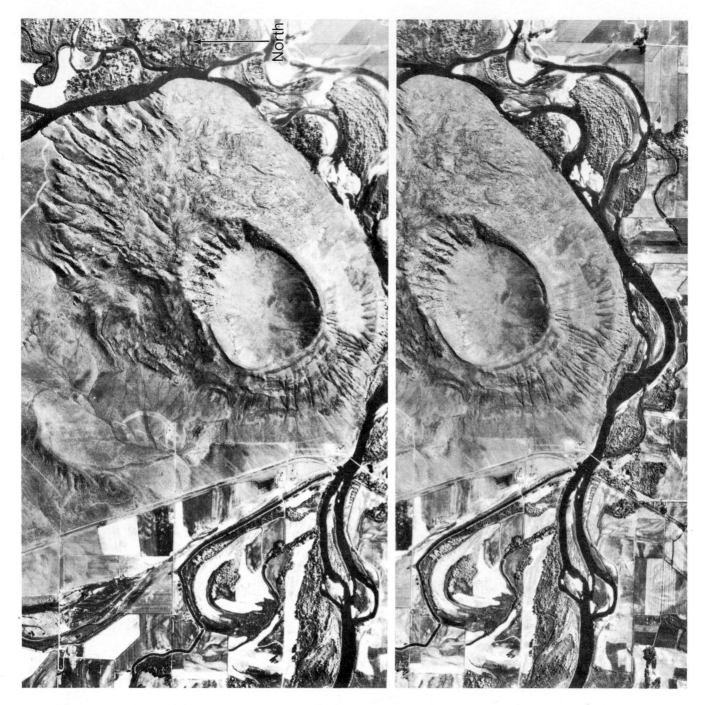

Figure 2.14   Stereopair of Menan Buttes, Idaho. The Snake River flows around the south side of the crater. (Courtesy of the U.S.G.S. Photos were taken on October 8, 1950.)

In both Parts 3 and 4 of this manual, several exercises will require interpretation of single aerial photos and stereopairs with reference to the terrain features and geologic phenomena shown on them. In preparation for those exercises, the paragraphs that follow will provide the student with some basic information on the interpretation of aerial photographs in terms of the recognition of common natural and man-made features.

Interpretation of Aerial Photographs

Air photos may be used to great advantage by geologists. Stereopairs are preferable, but single photos taken under good conditions of lighting and from proper altitudes reveal exactly what the human eye sees except for the third dimension, and even this can, with practice, be approximated from single photos.

Figure 2.15 Student using a lens stereoscope to view a stereopair. In actual practice, the nose is positioned in the slot between the two lenses in order to bring the eyes closer to the lenses.

The interpretation of air photos is an art acquired only after considerable experience in working with photos from many different areas. However, beginning students can comprehend air photos amazingly well if they have a few simple instructions to follow and some basic principles to guide them.

The greatest difficulty confronting an individual looking at an air photo for the first time is recognition of familiar features that are seen every day on the ground but become mysterious objects when seen from the air. It is necessary, therefore, to acquire the ability to recognize certain common features before one can expect to use an air photo as a geologic tool. Some of these common features are described in the paragraphs that follow and are illustrated in figure 2.16.

Vegetation

Vegetational cover accounts for a great many differences in pattern and shades of gray tone on air photos. Heavily forested areas are usually medium to dark gray, whereas grasslands show up in the lighter tones of gray. Planted field crops are extremely varied in tone, depending not only on the kind of crop but also on the stage of growth. Cultivated fields are usually rectangular in shape and appear either in dark gray or light gray, depending upon whether the fields have just been plowed or whether the crop is already in and growing.

Soil and Rock

Soil texture controls the soil moisture, which in turn controls the appearance of the soil on air photos. Wet clayey soils have a much darker shade of gray than do the dry sandy soils, which usually show up as light gray to almost white in arid regions. Different tones of gray that show up in the same field are due to different degrees of wetness of the soil, a condition usually related to topographic low (wet) and high (dry) areas.

Where bedrock crops out at the surface, air photos reveal differences due to lithology, texture, mineral composition, and structure of the rocks. The beginner cannot always discern between a sandstone and a limestone, for example, but should have less trouble recognizing the differences among the major rock groups (i.e., igneous, sedimentary, and metamorphic). Vegetational patterns on air photos commonly reflect the underlying bedrock, and this is helpful in tracing out a single rock unit of the photo. On the other hand, if a thick mantle of unconsolidated material, either residual or transported, covers the area, then all bedrock features may be partially obscured or entirely obliterated.

Air photos used in connection with field investigations on the ground are among the most valuable tools of the professional geologist. When direct field examination is not feasible, air photos provide even better information than one could acquire by flying over the area in person, especially if the photographs are available for stereoscopic study. For the student of elementary physical geology, aerial photographs are useful in that they show geologic features of the earth's surface that otherwise would be difficult to describe or impossible to illustrate.

Landsat False Color Images

The normal human eye can perceive a continuous spectrum from blue and green to yellow, orange, and red. Neither ultraviolet nor infrared radiation is visible to humans. The imaging technology used on the Landsat remote sensors records infrared radiation that, when combined with other wavelengths in the processing of the imagery data, results in a colored image in which the true colors are replaced by other colors. Hence the term, *false color image.*

In false color images, green vegetation shows up as various shades of red, deserts and other nonvegetated tracts are light gray to bluish gray, cities and large metropolitan areas are dark gray, and clear water areas are commonly black. Water bodies containing silt and other suspended sediments appear in various shades of light blue. Because

a. Sedimentary rocks uplifted to form dome (egg-shaped feature) in a semi-arid climatic zone. Stream valley with deciduous trees in valley bottom (black). Wyoming. Scale, 1:80,000.

b. Deeply incised river in flat lying sedimentary rocks. Badlands topography on either side of the river. Colorado. Scale, 1:80,000.

c. Small town in midwest. Various cultural features shown. Deciduous vegetation in town and along river. Missouri. Scale, 1:20,000.

d. Agricultural field patterns (light and dark variegated area) with undrained depressions forming ponds (white due to reflection of sun off water surface). Contour plowing evident. Texas. Scale, 1:63,360.

e. Crystalline rocks with sparse coniferous vegetation. River in deeply cut valley. Lake (black) at margin of photo. Joint pattern in rocks evident. Wyoming. Scale, 1:60,300.

f. Glacial moraine from continental glacier. Kame and kettle topography with numerous kettle lakes (black). Rectangular field patterns in variegated colors. North Dakota. Scale, 1:60,000.

Figure 2.16 Some examples of features visible on aerial photographs.

vegetational patterns reflect to some extent the underlying soil and rock types, the various shades and hues of red and other colors on a false color image define the different kinds of soil and rock types.

Landsat views do not cover a rectangular area due to the fact that the earth rotates as the satellite passes overhead. This phenomenon results in a rhombohedral "picture" such as the image shown in figure 2.17. North is generally toward the top margin of the image, but a true north-south line must be determined independently of the image margins. Where visible, agricultural fields are useful for this purpose because their boundaries are commonly oriented in north-south and east-west directions. If these are not present, other features such as a coastline or a major river can be compared with a map showing the same features to determine true north.

Figure 2.17 is a false color image of part of the Atlantic coast of the United States showing Long Island, the Hudson River, and a part of New Jersey. The Atlantic Ocean appears black, and the light-colored patches in New Jersey are cultivated fields. The brownish area in the southern part of the image is the forested coastal plain of New Jersey, and the dark area in the extreme northwest corner is part of the northern Appalachian Mountains. New York City and Newark, New Jersey, at the mouth of the Hudson River, are dark blue-gray. The light color of the narrow strips of coastal beaches off the southern shore of Long Island and off the coast of New Jersey reflects the fact that these features are sandy and generally lack heavy vegetation.

References

Avery, T. E., and Graydon, L. B. 1985. *Interpretation of aerial photographs.* 4th ed. Minneapolis: Burgess Publishing Co. 554 pp.

Colwell, R. N., ed. 1983. *Manual of remote sensing.* 2nd ed. Falls Church, Virginia: American Society of Photogrammetry and Remote Sensing. 2,724 pp.

Drury, S. A. 1987. *Image interpretation in geology.* London: Allen and Unwin. 243 pp.

Hyatt, E. 1988. *Keyguide to information sources in remote sensing.* London: Mansell Publishing Ltd. 274 pp.

Sabins, F. F. 1986. *Remote sensing: principles and interpretation.* 2d ed. Oxford: W. H. Freeman and Co. 592 pp.

Williams, R. S., and Carter, W. D. eds. 1976. *ERTS-1, a new window on our planet.* U.S.G.S. Professional Paper 929, Washington, D.C. 362 pp.

Figure 2.17   False color image of the Long Island–New Jersey area made from Landsat 2 on October 21, 1975. The long narrow strip of sand beach off the Southern shore of Long Island is about 80 miles long. (NASA ERTS image E–2272–14543, U.S.G.S. EROS Data Center, Sioux Falls, South Dakota 57198.)

Exercise 11. Introduction to Aerial Photograph Interpretation

1. Figure 2.18 is an aerial photograph of standard size. Identify the following features on the photograph. Where a topographic map symbol is available for the feature (refer to fig. 2.2), draw the symbol directly on the photograph of figure 2.18 at the place where the feature occurs. If no symbol exists, draw the outline of the feature on the map and label it accordingly. Use a red pencil in all cases except for water features, which should be shown in blue.
   a) Major road
   b) Railroad track
   c) River
   d) Landing strip
   e) Football field and track
   f) Small stream valley
   g) Bridge over river
   h) Overpass
   i) Golf course

2. The southeast corner of the golf course is also the SE corner of Section 12, T. 92 N., R. 52 W. The "T" intersection of the road at the south edge of the golf course about 4 inches to the west (left) on the photograph is at the SW corner of the SE¼ of the SE¼ of Section 11, T. 92 N., R. 52 W. Determine the R. F. and verbal scale of the photograph.

3. Figure 2.19 is a stereopair. Examine it with a stereoscope and answer the following questions while completing the instructions.
   a) Does the major stream channel contain any signs of the presence of water?
   b) Trace the drainage lines in blue pencil on the right-hand photograph using the proper symbol from figure 2.2.
   c) What is the dominant vegetation of the area?
   d) Are any man-made features visible on the stereopair?
   e) The area covered by the stereopair shows two relatively flat upland surfaces away from the stream channel, each of which lies at a general elevation that differs from the other. Draw the boundaries of these areas on one photograph while viewing the stereopair with a stereoscope. Use a red pencil. Mark the lower-lying area with the number *1*, the higher area with the number *2*. Extend these boundaries on the rest of the photograph.

4. Figure 2.13 covers part of the Delaware Map of figure 2.12. With red pencil, label the following features with the lowercase letter that appears before each name:
   a) Lake Superior
   b) Lake Bailey
   c) Agate Harbor
   d) Small lake in section 35
   e) Swamp area south of Mt. Lookout
   f) Mt. Lookout
   g) Trace the roads in sections 34, 35, and 36 that are shown by a dashed red line on the Delaware Map.

5. Trace the drainage system from the Delaware Map onto the aerial photo of figure 2.13. Use a blue pencil and label each creek or river that has a name.

6. Determine the scale of the aerial photo in figure 2.13.

7. Figure 2.7 is an aerial photograph of an area in central Iowa. Section 6 of T. 90 N., R. 37 W. is labeled on the photograph.
   a) Using the information provided, determine the scale of the photograph.
   b) Draw the township and range lines on the photograph using red pencil. Draw in the section lines with black pencil. Number all sections in black pencil. Along the photograph margins record the township and range designations in red pencil using the conventional system shown in figure 2.6.
   c) Note the relationship between the section lines and the road pattern. Compare this relationship with that on figure 2.18. Suggest possible reasons for the differences.
   d) A railroad track crosses the area from southeast to northwest. Use the appropriate map symbol to show this feature on the photograph. (Draw directly on the photograph with black pencil.)
   e) The rectangular areas covering most of the area are croplands. They appear in various shades of gray to nearly black on the photo. Explain the differences.

8. The scale of a vertical photograph depends on the focal length of the camera used and the height of the airplane taking the photo above the ground. The R. F. is the focal length of the camera divided by the height of the plane above the ground. Determine the scale of a photograph taken by a camera with a 9-inch focal length from an airplane flown at a height of 35,000 feet above the ground. (Be careful about units.)

Figure 2.18   Aerial photograph of area in South Dakota. (Photo No. VE-1JJ; taken on June 19, 1968.)

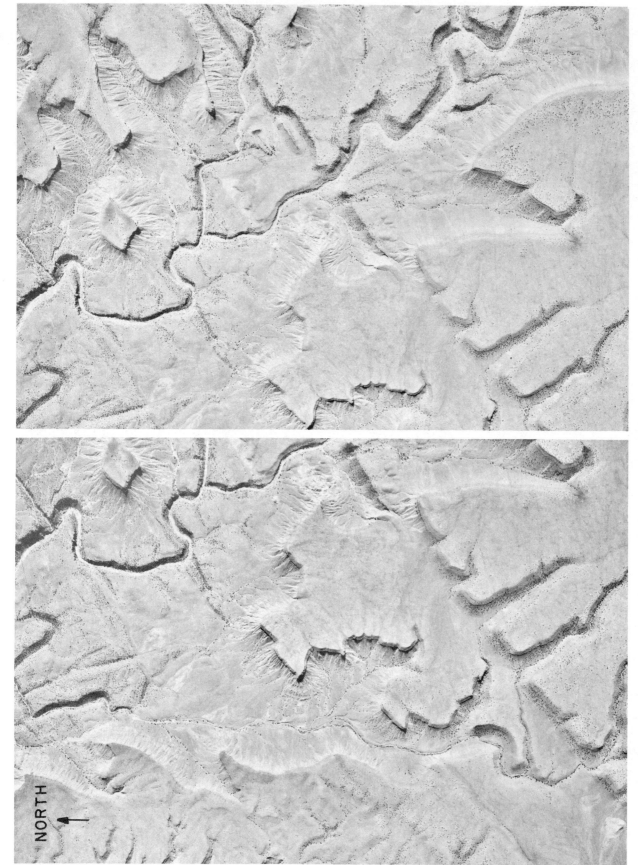

Figure 2.19 Aerial photograph stereopair, Utah, 1956. Scale, 1:20,000. (Photograph numbers GS-RR-17-43 and GS-RR-17-44.)

NORTH

Geologic Interpretation of Topographic Maps, Aerial Photographs, and Earth Satellite Images

Part 3

Background

In Part 2 of this manual we learned that the configuration of a landform is expressed on a topographic map by contour lines, and it is revealed on aerial photographs when two overlapping photos are viewed stereoscopically. Topographic maps and stereopairs can thus be used for the study and analysis of terrain features in terms of the geologic processes that produced them. Images from earth-orbiting satellites are additional tools useful in the study of landforms.

Every geologic process leaves some imprint on the part of the earth's surface over which it has been operative. These processes include the work of the major geologic agents such as wind, groundwater, running water, glaciers, waves, and volcanism. Each of these agents leaves its mark on the landscape in the form of one or more characteristic landforms.

The association of geologic agents with the origin of various landforms is a subdivision of geology called *geomorphology*. Geomorphologists have systematized the relationship of geologic processes to topographic forms into a body of knowledge that can be used in deciphering the origin of topographic features shown on a topographic map or seen in a stereopair. The body of knowledge that deals with the origin of landforms is presented in all basic textbooks that deal with physical geology.

It is assumed that the users of this manual will have become acquainted with the different geologic processes and their related landforms through reading appropriate chapters in a textbook on physical geology, and by listening to lectures in which the origin of landforms is presented. This is prerequisite to the understanding and successful completion of the exercises that are presented in this part of the manual.

General Instructions

The purpose of Exercises 12 through 19 is to acquaint the student with a variety of landforms and geologic principles associated with the geologic agents of wind, groundwater, running water, groundwater, glaciers, wind, waves, and volcanism. Topographic maps and profiles, aerial photographs, some of which are in the form of stereopairs, and other pertinent maps, satellite images, diagrams, and data are provided in Exercises 12 through 19 as the basic tools for learning the association between landform and geologic agent, or to establish a specific geologic principle.

The title of each exercise identifies the geologic agent that will be under consideration for that particular exercise. It is assumed that you are thoroughly conversant with and have a good understanding of the material presented in Part 2 of this manual on topographic maps, satellite images, and aerial photographs, including map and photo scales, contour lines, map symbols, and topographic profiles. The terms associated with the geologic processes covered in this part of the manual generally are defined in the background material for each exercise in which they are used. However, it is good practice to bring your textbook to the laboratory as a reference for unfamiliar or forgotten terms that crop up in the exercises.

All maps and photographs needed for completion of the exercises are included in the manual. Before proceeding with the questions or problems based on maps or photographs, note the scale and contour interval on the topographic maps, and the scale of the stereopairs. Some exercises require you to draw on the maps or photos with ordinary lead pencil or colored pencils. In these cases, make your initial lines very light so that if erasure is necessary, it can be done easily. Also, use sharp pencils to ensure accuracy, especially where the drawing of a topographic profile is a requirement of the exercise. Some students who suffer from eyestrain after prolonged study of maps may find an inexpensive magnifying glass helpful for completing the map exercises. The questions for each exercise should be answered in the order given because they are arranged in a more or less logical sequence.

Geologic Work of Running Water

Stream Gradient and Base Level

Landforms produced by running water are produced by stream erosion, stream deposition, or a combination of erosion and deposition. The *gradient* of a stream is the slope of the stream bed, or surface of the stream in larger rivers, along its course. A *longitudinal profile* of a stream shows that the gradient decreases in a downstream direction (fig. 3.1*A*). A stream tends to erode its channel bottom or bed in the upper or headward reaches, and deposit sediment or erode its channel walls in the lower reaches.

Base level is the lowest level to which a stream can erode its bed. The base level for streams that flow to the ocean is sea level, but other temporary base levels may exist along the stream course. These include lakes created by dams (fig. 3.1*B*).

The surface of a lake behind a dam constitutes a new base level for the segment of the stream above the dam.

Sediment is deposited where the stream enters the lake because the stream gradient has been reduced. Below the dam, a new profile is formed because the water discharged through the gates of the dam is relatively free of sediment, thereby allowing the stream to erode its bed to a deeper level than before the dam was built.

Stream Gradients and Drainage Divides

A stream together with its tributaries is called a *drainage system.* Contiguous drainage systems are separated by a *drainage divide,* an imaginary line connecting points of highest elevations between the two systems (fig. 3.2*A*).

The erosive power of a stream is a function of the stream gradient. Streams with higher gradients are more potent in eroding their beds than streams with lower gradients. If adjacent drainage systems have comparable gradients in their upper reaches, the divide between the

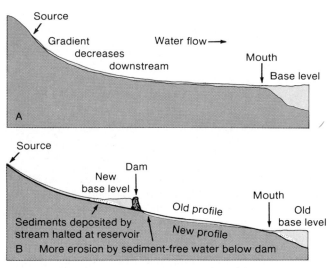

Figure 3.1 (*A*) Longitudinal profile of a river showing a gradual decrease in gradient downstream. Base level is the lowest to which a stream can erode its bed. (*B*) Longitudinal river profile that has been interrupted by a dam. The old, pre-dam base level has been replaced by a new one. (From Carla W. Montgomery, *Physical Geology,* 2d. ed. Copyright © 1990 Wm. C. Brown Publishers, Dubuque, Iowa. All Rights Reserved. Reprinted by permission.)

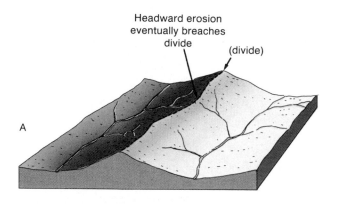

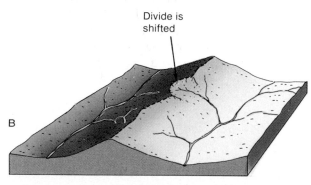

Figure 3.2 (*A*) A divide between two opposing drainage systems is attacked by headward erosion. (*B*) The divide is shifted toward the drainage system whose streams have the smaller gradient of the two. (From Carla W. Montgomery, *Physical Geology.* Copyright © 1987 Wm. C. Brown Publishers, Dubuque, Iowa. All Rights Reserved. Reprinted by permission.)

two systems will remain more or less constant. If, however, the gradients of the streams in the two systems differ appreciably, the divide will move over time toward the system with the smaller gradient (fig. 3.2B).

Pediments and Alluvial Fans

These two landforms have similar but not identical topographic expressions. An *alluvial fan* is a depositional feature produced where the gradient of a stream changes from steep to shallow as it emerges from a mountainous terrain. At that point the coarse sediments carried by the fast-flowing stream in the mountain are deposited in the form of a fan-shaped apron at the mountain front where the gradient decreases abruptly.

A *pediment,* on the other hand, is a gently sloping *erosional surface* covered with a veneer of coarse sediment. Pediments are common in arid regions. They form as the weathered mountain front recedes by attack from heavy but infrequent rainfalls. Below the coarse sedimentary cover in transit lies the planed-off bedrock that was formerly part of the mountainous terrain. One description of a pediment is that it is the end product of a mountain range consumed by weathering and stream erosion. A pediment may contain remnants of the mountainous terrain standing as isolated bedrock knobs above the pediment surface.

Meandering Rivers and Oxbow Lakes

In the lower reaches of a stream's course where the gradient is low, the stream's erosive action is directed toward the channel walls. This process of *lateral erosion* produces a flat valley floor called a *floodplain* (fig. 3.3). A floodplain is not produced by flood stages of the river but rather by constant shifting of the stream channel through lateral erosion across the valley floor. A floodplain does become inundated when the stream channel is unable to accommodate the volume of spring runoff or of heavy rains in the drainage basin at any time of the year.

The circuitous course of a stream flowing across its floodplain is called a *meandering course,* and the individual bends are called *meanders.* When a meander loop is isolated from the main channel by erosion of the narrow neck of land between the upstream and downstream segments of the meander loop, an *oxbow lake* is formed (fig. 3.4). With passage of time, an oxbow lake gradually becomes filled with sediment brought in by flood waters, and eventually is reduced to a swampy scar on the floodplain as testimony to its former existence as an active channel of the river.

Floodplains are attractive sites for agriculture and land development. To protect structures built on them such as roads, buildings, and airports, *levees* are built along both sides of a meandering river to contain flood waters.

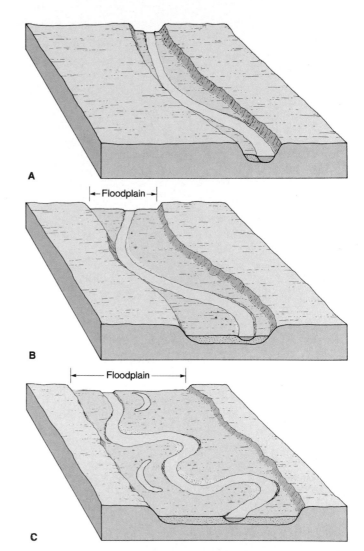

Figure 3.3 The evolution of a floodplain. (*A*) River widens the valley floor by lateral erosion. (*B*) Continued lateral erosion widens the valley floor further, and the river course becomes more meandering. (*C*) The river's course develops meander loops, which become oxbow lakes. (From Carla W. Montgomery, *Physical Geology,* 2d. ed. Copyright © 1990 Wm. C. Brown Publishers, Dubuque, Iowa. All Rights Reserved. Reprinted by permission.)

The lower Mississippi River is a classic example of a meandering stream flowing across a broad floodplain. Hundreds of miles of levees have been constructed by the U.S. Army Corps of Engineers to contain flood waters. Additionally, the Corps of Engineers have cut off meanders from the main channel through the excavation of channels across meander necks. These artificial *cutoffs* shortened the river course in deference to the barge traffic plying the river with cargoes of various sorts.

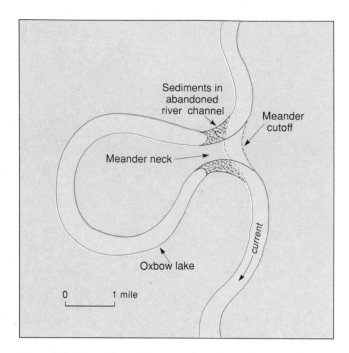

Sediments in
abandoned
river channel

Meander
cutoff

Meander neck

current

Oxbow lake

0   1 mile

Figure 3.4   Map diagram showing a former river meander that has been severed from the river by a meander cutoff that flows across the meander neck. The ends of the old meander have been filled with sediment, thereby forming a closed depression called an oxbow lake.

The Mississippi River forms the boundary between many states from Minnesota to its mouth in Louisiana, where it enters the Gulf of Mexico. The states of Mississippi and Arkansas generally lie east and west of the river, respectively. Some small segments of each of these states, however, lie on the *opposite* side of the river. The reason for this is based on the meandering nature of the river and the need to fix the boundary so that the shifting course of the river does not also shift the state boundary. To reduce interstate conflicts that might arise through a constantly shifting boundary, the United States Supreme Court ruled in 1820 that where a stream forms the boundary between states, and the channel of the stream changes by the "natural and gradual processes known as erosion and accretion, the boundary follows the varying course of the stream; while if the stream from any cause, natural or artificial, suddenly leaves its old bed and forms a new one, by the process known as avulsion, the resulting change of channel works no change of boundary, which remains in the middle of the old channel, although no water may be flowing in it."*

*Van Zandt, Franklin K. 1976. *Boundaries of the United States and the Several States.* U.S.G.S. Professional Paper 909, p. 4.

Exercise 12A. Stream Gradients and Base Level

A segment of the Missouri River flows across the Portage Map in figure 3.5.

1. What is the direction of flow of the Missouri River?

2. Three dams are located along the course of the Missouri River. On the basis of man-made installations associated with each dam, what purposes do the dams serve?

3. How many new base levels have been created by these dams?

4. In the stretch of the Missouri River between the Rainbow and Ryan dams, is the water depth greater near the Rainbow Dam or the Ryan Dam?

5. In the same stretch of river, state whether the riverbed is subject to erosion or deposition at the following sites:
   a) Immediately downstream from the Rainbow Dam.
   b) Immediately upstream from the Ryan Dam.

6. If the Ryan Dam were destroyed by an earthquake, describe the environmental effects on
   a) The channel of the Missouri River between Ryan Dam and Rainbow Dam.
   b) Morony Dam.

7. Beginning about 1 mile downstream from Rainbow Dam, a series of contour lines cross the Missouri River. In a distance of 1.7 river miles, the river drops 60 feet. On the basis of this information, what is the gradient of the river along this stretch? State your answer in feet per mile.

8. The Missouri River flows a total distance of 2,500 miles from its headwaters at 14,000 feet above sea level to its confluence with the Mississippi River at 410 feet above sea level. What is the average gradient of the Missouri River from source to mouth? Why does the average gradient differ from the gradient determined in question 7?

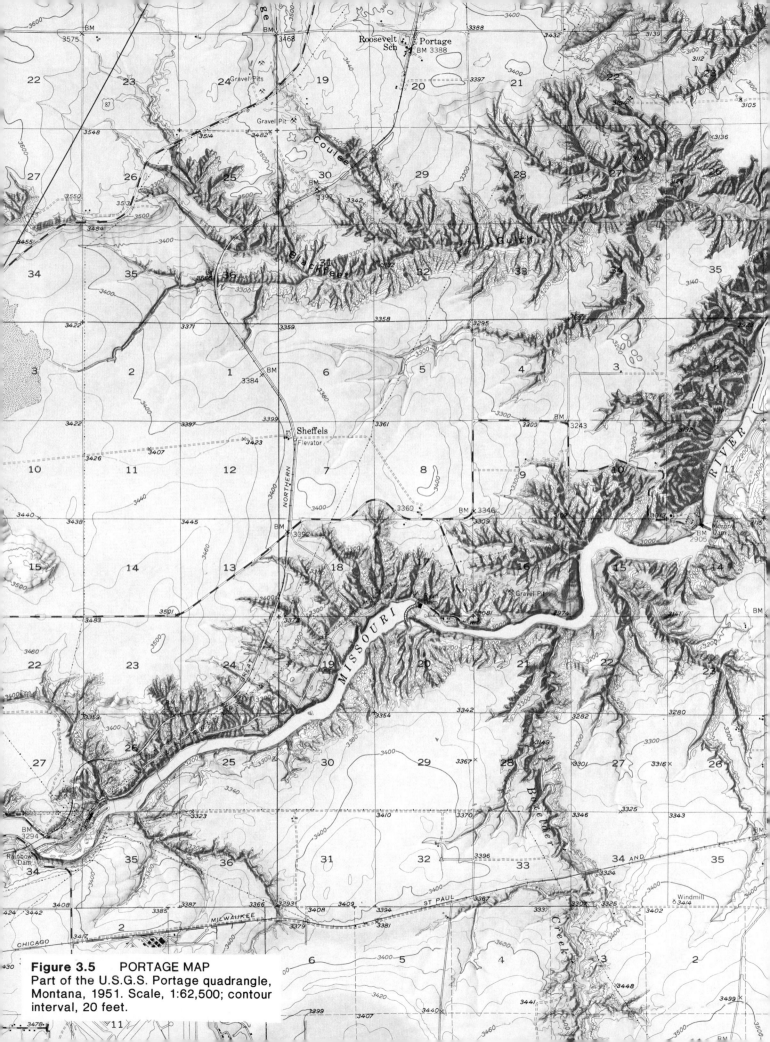

Figure 3.5   PORTAGE MAP
Part of the U.S.G.S. Portage quadrangle,
Montana, 1951. Scale, 1:62,500; contour
interval, 20 feet.

Exercise 12B. Stream Gradients and Drainage Divides

The Promontory Butte Map (fig. 3.6) shows two stream systems flowing north and south of the Mogollon Rim (pronounced Mo-gee-yone). The topography produced by these two systems is more rugged south of the rim than north of it. In this exercise we will examine the reason why this difference exists.

1. Trace in red pencil the divide between the headward regions of the two stream systems. Use a soft black pencil first, then trace it in red when you are satisfied with its position.

2. What man-made feature is *roughly* coincident with the divide?

3. Does the Mogollon Rim lie on the divide or north or south of it?

4. Table 3.1 lists several streams north (*A*) and south (*B*) of the Mogollon Rim. For each stream, the differences in elevation of two contour lines that cross the stream a few miles apart are recorded, and the map distance between them is shown.
   a) For each stream segment listed in table 3.1, calculate its gradient in feet per mile.
   b) Determine the average gradient for the streams north and south of the Rim.

5. If the erosive potential of a stream is proportional to its gradient, which system, *A* or *B*, should have the greatest erosive power?

6. Will the divide between the two systems move generally toward the north, toward the south, or remain more or less fixed with the passage of time? Explain your answer.

7. Sketch a series of maps that show the progressive changes that will occur around Promontory Butte, especially at its juncture with the Mogollon Mesa.

Table 3.1 Matrix for Determining Stream Gradients, (A) North of the Mogollon Rim, and (B) South of the Mogollon Rim.

A. Stream Gradients, North of Mogollon Rim.

Name of Stream	Length in Miles	Diff. in Elev. in Feet	Gradient in ft/mi
West Leonard Canyon	2.6	250	
Middle Leonard Canyon	2.4	400	
East Leonard Canyon	1.9	250	
Turkey Canyon	1.9	250	
Beaver Canyon	3.7	300	
Bear Canyon	1.6	200	
	Average Gradient		

B. Stream Gradients, South of Mogollon Rim

Name of Stream	Length in Miles	Diff. in Elev. in Feet	Gradient in ft/mi
Big Canyon Creek	3.7	1,500	
Dick Williams Creek	1.8	1,150	
Horton Creek	3.0	1,000	
Doubtful Canyon	2.4	1,250	
Spring Creek	2.2	1,250	
See Canyon	2.2	700	
	Average Gradient		

Figure 3.6 PROMONTORY BUTTE MAP
Part of the U.S.G.S. Promontory Butte
quadrangle, Arizona, 1951. Scale, 1:62,500;
contour interval, 50 feet.

Exercise 12C. Pediments and Alluvial Fans

The Antelope Peak Map (fig. 3.7) shows a well-developed pediment, and the Ennis Map (fig. 3.8) depicts an excellent alluvial fan.

1. On the grid of figure 3.9, draw a topographic profile of each of these features with the beginning and end points of the profile as described below. Begin each profile by plotting the *highest* elevation on the vertical axis at mile zero.

   a) Line of profile for the Antelope Peak pediment: the starting point is at the SW section corner of Section 35, T. 6 S., R. 2 E., at an elevation of 1,694 feet. The profile extends as a straight line for a little more than 6 miles in a northeasterly direction, passing through the SW section corner of Section 6, T. 6 S., R. 3 E., at an elevation of 1,320 feet, and ends at the 1,300-foot contour line about one-half mile northeast of the previously described point. Draw the line of profile on the Antelope Peak Map with a sharp pencil.

   b) Line of profile for the Cedar Creek Alluvial Fan, Ennis, Montana: the starting point is at the northeast section corner of Section 21 in the southeast part of the Ennis Map area (Section 21 is the one in which the Lawton Ranch is located). The elevation at the starting point (i.e., mile-zero) is 6,080 feet above sea level. From this point, the line of profile extends as a straight line in a northwesterly direction for about 4 miles, passing through the northwest section corner of Section 7 (elevation 5,245 ft), and ending at the 5,200-foot contour line. (In drawing this profile, it will be more convenient to turn the map upside down so that the starting point of the profile will be at your upper left.) Draw the line of profile on the Ennis Map with a sharp pencil.

2. Both the pediment and the alluvial fan are produced by running water. Describe each profile in terms of its shape (i.e., concave, convex, or straight) and gradient in feet per mile.

3. Inasmuch as both the pediment and the alluvial fan are produced by the action of running water, suggest a reason or reasons why the two profiles are different.

4. The uniform regularity of contour spacing on the pediment of the Antelope Peak Map is interrupted by several isolated hills that rise 100 to 200 feet above the level of the pediment between Mesquite Road and Highway 84. What is the origin of these features?

Figure 3.7 ANTELOPE PEAK MAP
Part of the U.S.G.S. Antelope Peak
quadrangle, Arizona, 1963. Scale, 1:62,500;
contour interval, 25 feet.

Figure 3.8 ENNIS MAP
Part of the U.S.G.S. Ennis quadrangle, Montana, 1949. Scale, 1:62,500; contour interval, 40 feet.

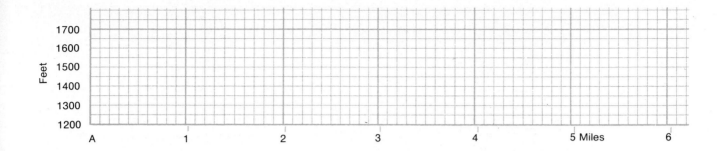

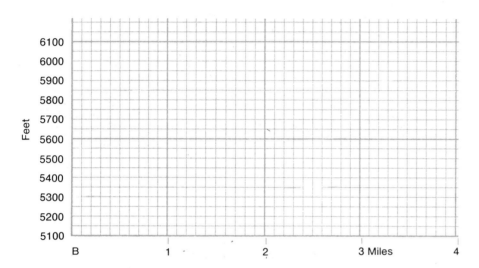

Figure 3.9 (*A*) Grid for drawing pediment profile based on the Antelope Peak Map. (See Exercise 12C–1a for location.) (*B*) Grid for drawing a profile of the Cedar Creek Alluvial Fan, Ennis Map. (See Exercise 12C–1b for location.)

Exercise 12D. Meandering Rivers

A close examination of the Greenwood Map of figure 3.10 reveals that the boundary between Arkansas and Mississippi does not follow the present course of the Mississippi River. The course of the Mississippi River when the boundary was fixed in 1909 by an act of Congress will be referred to as the *Boundary Course,* and the present course will be called the *Modern Course.*

1. Trace the Boundary Course in red pencil and the Modern Course in blue pencil.

2. Between the north and south boundaries of the Greenwood Map, the Boundary Course is 136 miles long and the Modern Course is 65 miles long.
   a) Why is the Modern Course shorter than the Boundary Course?
   b) Express as a percent the amount of shortening of the Modern Course compared to the Boundary Course.
   c) What is the impact of the shortening on the gradient of the segment of the river shown on the Greenwood Map?

3. Why is it impossible to determine the gradient of the Mississippi River for either the Boundary Course or the Modern Course from information on the Greenwood Map?

4. The Arkansas-Mississippi boundary east of Lake Chicot is labeled "indefinite." Given the 1820 Supreme Court decision, what conditions must exist along this stretch of the river to justify the labeling of the interstate boundary as indefinite?

5. Construction of artificial cutoffs by the Corps of Engineers in the 1930s created several oxbow lakes. These lakes or their remnants are, from north to south, Lake Whittington, Lake Paradise, unnamed lake west of Tarpley Cut-off, Lake Ferguson, and Lake Lee. Write the letters ARK or MISS in the areas circumscribed by these oxbows to indicate which state has jurisdiction over them. Did Mississippi or Arkansas gain more land when the permanent boundary between them was established?

6. Are the lakes mentioned in question 5 likely or unlikely to be inundated during flood stages of the Mississippi River? Why?

7. Is Lake Chicot older or younger than the Boundary Course?

Figure 3.10   GREENWOOD MAP
Part of the U.S.G.S. Greenwood, Miss.–Ark.–La.
Map, 1953, revised in 1979. Scale, 1:250,000;
contour interval, 50 feet with supplementary
contours at 25-foot intervals.

Groundwater Movement, Groundwater Pollution, and Groundwater as a Geologic Agent

Groundwater

Groundwater is the water that occurs beneath the surface of the earth in the *zone of saturation*. The top of the zone of saturation is the *water table*. Groundwater originates as rain that percolates downward through porous soil and rock until it reaches the water table. The water table lies at variable depths below the surface of the earth.

Water Table Contours

The water table is a planar feature and its configuration can therefore be defined by contour lines in the same way that the configuration of the earth's surface can be defined by contour lines on a topographic map. Contour lines on the water table can be drawn if enough points of known elevation on the water table are available for plotting on a base map. These points are assembled from water wells and other places of known elevation where the water table intersects the earth's surface, such as a stream or a lake.

Flow Lines

Groundwater moves under the influence of gravity through porous rock or unindurated sand and gravel, but the movement is very slow compared to flowing water in a river. Groundwater moves along flow lines. A *flow line* is a path followed by a water molecule from the time it enters the zone of saturation until it reaches a lake or stream where it becomes surface water.

Figure 3.11 is a map of a hypothetical area underlain by clean sand showing the relationship of the water table contours and flow lines to a permanent surface stream flowing south. For example, a water molecule at point *a* will follow the path of the flow line until it enters the stream at *a'*. The same relationship holds true for water molecules at *b, c, d,* and *e*. They follow the flow lines as they move down the slope of the water table at right angles to the water table contours until they enter the stream at points *b', c', d',* and *e'*, respectively.

Flow lines can converge or diverge, but they cannot cross each other. Moreover, as can be deduced from figure 3.11, groundwater from the west side of the stream cannot move across the stream to be commingled with groundwater on the east side of the stream, and vice versa, because the stream intercepts the flow of groundwater from both sides. From this simple analysis, it follows that any pollutant introduced into the zone of saturation will be carried by the flow of groundwater along flow lines until it eventually is discharged into a lake or stream.

Conversely, if a pollutant were introduced into the stream of figure 3.11 at points *a'*, it would not enter the water table but would flow down the stream channel in the direction of *b'*. This relationship between groundwater flow and stream flow holds only for the case of an *effluent stream*, which is *a stream fed by groundwater;* that is, the groundwater is discharged into the stream channel. An *influent stream,* on the other hand, is one in which the *groundwater flows away from the stream channel.* In such cases, pollutants dumped directly into the stream will move into the zone of saturation. Influent streams also are commonly *intermittent streams,* streams that flow only during certain times of the year when rainfall is sufficient to supply surface runoff directly to them. With these basic principles in mind, we will now apply them to the following exercises.

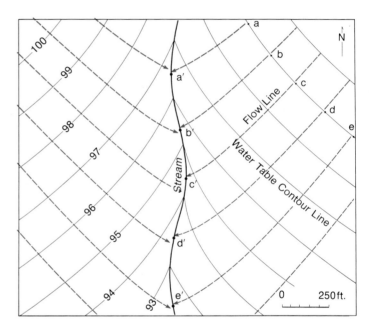

Figure 3.11   Map of a hypothetical area underlain by a well-sorted coarse sand showing a south-flowing permanent stream and contours on the water table. The ground surface is roughly the same shape as the water table but about 5 to 10 feet higher. The contour interval of the water table contours is 1 foot. Flow lines are shown in dashed colored lines. (See text for further explanation.)

Karst Topography

Groundwater slowly dissolves such rocks as limestone, dolomite, and rock salt. This process of *groundwater solution* forms caves and connecting passageways. Ultimately, the roofs of these underground cavities collapse, leaving surface depressions called *sinks* or *sinkholes*.

A terrain marked by many sinks is called *karst topography,* a name derived from a limestone region along the Dalmatian coast of Yugoslavia. In the United States, two areas where karst topography is well developed lie in Florida and Kentucky. The Lake District in central Florida consists of thousands of water-filled sinks, and the famous Mammoth Cave of Kentucky is part of a system of interconnecting caverns and passages formed in limestone.

As a Karst terrain evolves, surface streams are replaced by a system of interconnected underground passageways that function essentially as a subterranean drainage system. Karst terrain is therefore generally devoid of surface streams.

Exercise 13A. Water Table Contours

The Ashby Map (fig. 3.12) covers part of the Sand Hills of western Nebraska. The Sand Hills are formed from ancient sand dunes that are now more or less stabilized by surface vegetation consisting mainly of native grasses, an environment that makes this an excellent area for the grazing of cattle.

Sand dunes are composed of windblown sand that is very porous. Rain striking the surface in the Sand Hills very quickly percolates to the zone of saturation where it becomes groundwater. The water table in the area covered by the Ashby Map is quite shallow (i.e., close to the surface) in the interdune "valleys," which explains the many lakes and marshes that occur there.

The numbers imprinted on some of the lakes in this map area are elevations of their water surfaces. For example, the water surface elevation of Castle Lake is 3,767 feet above sea level. Assuming that these elevations are points on the water table, it is possible to draw a rough approximation of the water table contours. Some lakes have no elevations marked on the map. In these cases, the lake elevation can be estimated by noting the elevation of the contour nearest to the lake shore. Careful inspection of the map reveals that the lowest known lake elevation is North Twin Lake in the northeast corner of the map (3,737 feet above sea level), and that the highest lake elevation is Melvin Lake in the northwest corner of the map area (3,798 feet above sea level). From this relationship we can deduce that the water table must be sloping downward in an easterly direction. Therefore, the contour lines on the water table must extend at right angles to the direction of slope of the water table, or in a general northsouth direction.

1. With a soft pencil (easily erasable) sketch in the water table contours for the entire map area. Use a C.I. of 10 feet. Start with the 3,800-foot contour line that begins just west of Melvin Lake and extends in a southerly direction until it passes just west of Hibbler Lake (about 2 miles northeast of the town of Ashby) with an elevation of 3,796 feet above sea level. Draw all water table contours on the map. For best results work first on the contours in the south-central part of the map area where more lakes are present, hence more control points are available. (Be sure that each water table contour is labeled.)

2. Locate the point in Section 20 (south half of the map) marked with a brown *x* and the numerals 3,862, which is the surface elevation at that point. On the basis of the water table contours that you have drawn on the map, estimate the depth of the water table at this point.

3. Why are there so few surface streams in this area?

3. Sand good aquifer - very permeable

Reference

Sniegocki, R. T. 1959. Geologic and groundwater reconnaissance of the Loup River Drainage basin Nebraska, *U.S.G.S. Water-Supply Paper 1493*. Washington, D.C.: U.S. Government Printing Office, 106 pp.

Figure 3.12   ASHBY MAP
Part of the U.S.G.S. Ashby quadrangle,
Nebraska, 1948. Scale, 1:62,500; contour
interval, 20 feet.

Exercise 13B. Groundwater Pollution

Figure 3.13 is the map of a hypothetical area underlain by well-sorted coarse sand crossed by a permanent stream, Clear Creek, that flows in a southeasterly direction. The ground surface is gently sloping to the southeast, and the shallow water table is defined by the water table contours. The location of a dump is shown on the map. The dump is privately owned and operated by a small company that hauls trash and garbage for residents of a nearby small town. The dump is an excavated pit, the bottom of which lies just above the water table.

The Jones property lies southeast of the dump, and its west property line borders on Clear Creek. Mr. Jones owns horses that he keeps in the barn and corral part of the time. The rest of the time the horses graze on the property and occasionally drink the water from Clear Creek.

The Smith estate lies west of Clear Creek and also has frontage on the creek. Both Jones and Smith derive their domestic water supplies from wells that penetrate the shallow water table. To assure themselves that the water was suitable for human consumption, Jones and Smith had well-water samples analyzed by the county health department at the time their wells were completed. Both Smith and Jones owned their respective properties for many years prior to the establishment of the dump, and have enjoyed a potable water until recently.

Recently, Jone's well water began to deteriorate in quality. This was verified when Jones had his water tested again at the county health department. Jones attributed this to pollution of the groundwater from leaching of domestic waste deposited in the dump. In talking with his neighbor, Smith, Jones suggested that the two of them should file suit against the owner of the dump and obtain a court injunction that would require cessation of all further dumping.

So Smith had water from his well tested again and found that it had not changed in quality since the tests conducted prior to the creation of the dump. Yet stream samples along the stretch that forms the boundary between the Jones and Smith properties were also analyzed by the county health department and were found to be contaminated by materials similar to those found in the recent water samples from the Jones well. This was sufficient evidence to convince Smith that he ought to join in the suit with Jones against the dump operator. Smith reasoned that if the creek was contaminated with the same materials found in the Jones well, it would be only a matter of time until his well would also be polluted.

At that point, Jones and Smith hired an attorney to file the suit. He sought the advice of a geologist at a nearby university who had access to publications of the U.S.G.S. in the school's library. There he discovered a general geology report of the geology of the area. A section of the report on groundwater contained a map showing water table contours based on static water levels in other wells not shown in figure 3.13. The geologist transferred those contours to a map he was preparing for the lawyer. This map is figure 3.13.

1. On the map of figure 3.13, several dots are shown along the 800-foot water table contour line. Assume that these are the points of intersection of flow lines with the water table contours. Using the relationship of flow lines to water table contours as shown in figure 3.11, sketch a network of flow lines on figure 3.13. One flow line should pass through each of the dots on the 800-foot contour line. (Use a soft black pencil because you may have to erase several times before you are satisfied with your results.) Extend the flow lines across the entire map area so that the movement of groundwater can be ascertained.

2. On the basis of the flow line network that you have constructed on figure 3.13, answer the following questions:
   a) Is there reasonable evidence to conclude that seepage from the dump has contaminated the Jones well? Explain.
   b) Is there reasonable evidence that the stretch of Clear Creek adjoining the Jones and Smith properties has been contaminated by seepage from the dump? Explain.
   c) Is there reasonable evidence that the Smith well will be contaminated by seepage from the dump at some time in the future? Explain.

3. Is it possible that the animal waste in the corral on the Jones property is responsible for polluting the Jones well, any part of Clear Creek, or the Smith well, eventually? Explain.

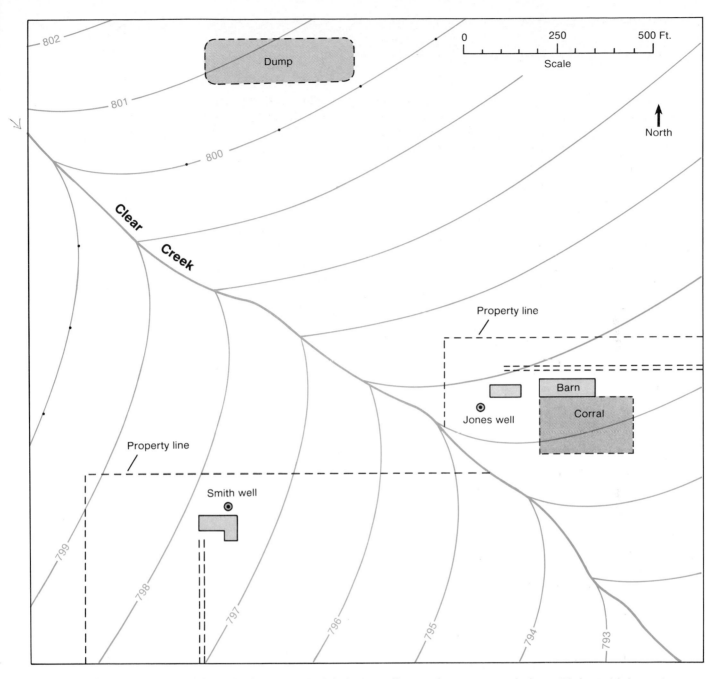

Figure 3.13   Map of a hypothetical area underlain by a well-sorted coarse sand about 50 feet thick. Clear Creek flows to the southeast. The water table contour interval is 1 foot. The water table lies about 8 to 10 feet below the ground surface, except near Clear Creek where the water table becomes shallower until it intersects the creek. The dump is an excavated pit, the bottom of which does not quite reach the water table.

Exercise 14A. Sinkholes and Karst Topography

Most of the sinks in the area covered by the Putnam Hall Map (fig. 3.14) are identified by closed contours (see p. 58 to review closed contours). Some of the sinks contain lakes. The stippled purple areas show dry lake beds determined from aerial photos in 1970.

1. Assume that the surfaces of the lakes correspond to points on the water table, and that the elevation of a lake surface lies 5 feet *below* the contour nearest to the lake shore. Write the surface elevation on each lake on the Putnam Hall Map.

2. Based on the lake surface elevations determined in question 1, is the water table perceptibly sloping or essentially flat?

3. Why do some sinks contain lakes and others do not?

4. A point marked with an X in Section 18 has a surface elevation of 157 feet. How deep would a water well have to be drilled at this point to intersect the water table?

5. Why are there no surface streams in the area?

6. The areas identified by purple stippling were added in the 1970 revision of the map. With what features are these areas associated, and what changes might they represent between 1949 and 1970?

Figure 3.14   PUTNAM HALL MAP
Part of the U.S.G.S. Putnam Hall quadrangle, Florida, 1949, revised 1970. Scale 1:24,000; contour interval, 10 feet.

Exercise 14B. Evolution of a Karst Terrain

The area covered by the Mammoth Cave Map (fig. 3.16) is a good example of the effect of underlying strata on the topography. A cursory inspection of the map reveals that the southern one-third of the map area is markedly different than the northern two-thirds from a topographic point of view. The reason for this is readily apparent from the geologic cross-section in figure 3.15. This north-south cross-section passes through the main road intersection in the town of Cedar Spring near the center of the map, and its horizontal scale is the same as the scale of the map. The pronounced change in topography on the map is marked by the Dripping Spring Escarpment lying just north of and roughly parallel to the Louisville and Nashville Road.

1. Draw the line of the geologic cross-section on the map in black pencil.

2. Generally speaking, the area north of the Dripping Spring Escarpment and *west* of the line of the cross-section is characterized by an integrated stream system. Use a blue pencil to trace the drainage lines occupied by permanent and intermittent streams. Use the correct map symbol for each. Take care not to extend your blue lines beyond the limits of the streams as they are shown on the map. What is the dominant bedrock in the area drained by the stream system?

3. Why does the stream flowing north into Double Sink end so abruptly there?

4. Examine the topography *east* of the line of cross-section and north of Dripping Spring Escarpment. A number of valleys such as Cedar Spring Valley, Woolsey Hollow, and Owens Valley resemble stream-cut valleys, but there are no streams flowing in them. Account for the fact that the topography of this area resembles a stream-dissected terrain even though no streams occupy the existing valleys.

5. In making a comparison of the terrains east and west of the line of the cross-section (fig. 3.16), which of the following statements is most likely the correct one?
   a) The area *west* of the line of cross-section will eventually resemble the area *east* of the cross-section as the streams cut deeper into the sandstone and encounter the underlying limestone.
   b) The area *east* of the line of cross-section will eventually resemble the area *west* of the cross-section as the overlying sandstone is further eroded.

6. Assume that the sandstone formation in figure 3.15 had an original thickness of 100 feet. On figure 3.15, draw the upper and lower contacts of the sandstone formation as it existed at some time in the geologic past before stream erosion began stripping it away.

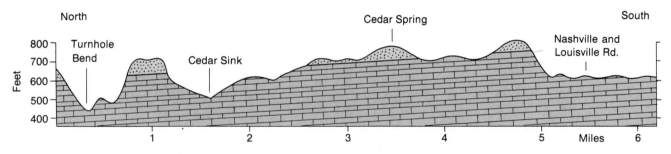

Figure 3.15   North-south geological cross-section from Turnhole Bend on the Green River to a point about three-fourths of a mile south of the Louisville and Nashville Road, Mammoth Cave Map. Geology is simplified from *The Geology of the Rhoda Quadrangle,* Kentucky, U.S.G.S. Map GQ-219 (1963). Vertical scale exaggerated about 10 times.

Figure 3.16   MAMMOTH CAVE MAP
Part of the U.S.G.S. Mammoth Cave
quadrangle, Kentucky, 1955. Scale, 1:62,500;
contour interval, 20 feet.

Glaciers and Glacial Geology

A *glacier* is a mass of flowing land ice derived from snow-fall. The two major types of glaciers are alpine or valley glaciers and continental glaciers or ice sheets. An *alpine glacier* is one that is confined to a valley and is literally a river of ice. An *ice sheet* or *continental glacier* covers an area of continental proportions and is not confined to a single valley. Landforms produced by both alpine and continental glaciers are distinctive features that can be recognized on aerial photographs and topographic maps.

Glaciology is the study of snow and ice. Alpine or valley glaciers occur in mountain valleys where adequate snowfall sustains them. Most of this snow falls during the winter and covers the entire glacier. During the ensuing summer, some or all of the previous winter snowfall is melted and is discharged from the glacier in meltwater streams. The snow that remains at the end of the summer melt season gradually changes to ice and becomes part of the glacier.

Wastage and Accumulation

The 12-month period of winter snow accumulation (accumulation) and summer melting (wastage) is known as the *budget year of a glacier.* In either hemisphere, the budget year begins at the end of the melt season just before the first snows of the winter, and ends about 12 months later at the end of the wastage season.

During the wastage months of the budget year, the previous winter's accumulation is partly removed from the upper reaches of the glacier. The snow that remains there lies in the *zone of accumulation.* Over the lower reaches of the glacier, the previous winter's snowfall is completely removed, and some of the underlying glacier ice is also melted. The area of the glacier that suffers wastage of both snow and ice is called the *wastage zone.* The line that separates the accumulation and wastage zones for a given budget year is the *annual snow line.* This is shown by a dashed line on the photograph of the Snow Glacier in figure 3.17 that was taken at the end of the wastage season. The white area above the snowline of the Snow Glacier is the residue of the previous winter's snowfall that survived the summer melting. On the lower reaches of the Snow Glacier, all of the winter snow was melted so that bare glacier ice is exposed. Some of this ice has also been lost by summer melting. The streams flowing from the glacier terminus in the lower right-hand corner of the photograph are fed by melted snow and ice from the glacier.

Glacier Mass Balance

It is possible to measure both wastage and accumulation for a given glacier during a budget year. The relationship between the annual wastage and accumulation over a period of years describes the general health of a glacier. For example, if wastage exceeds accumulation for a period of 10 to 20 years, the total mass of the glacier will decrease, a condition usually reflected in the retreat of the glacier terminus and a reduction in thickness of the glacier. If accumulation exceeds wastage for several years, the glacier terminus advances and the glacier thickens.

A comparison between accumulation and wastage for a given budget year yields the *net mass balance* of the glacier. Net mass balance is expressed numerically in feet or meters of water equivalent, and represents a hypothetical layer of water determined by many measured thicknesses of columns of snow or ice on the glacier's surface. If a glacier gains in mass during a budget year (i.e., accumulation exceeds wastage), the glacier is said to have a *positive net mass balance.* If the glacier loses mass during the budget year, it has a *negative net mass balance.*

Alpine Glaciers

An alpine glacier erodes the floor and walls of the valley through which it flows. Furthermore, it functions as a conveyor belt that transports debris from the valley floor and walls to the glacier terminus where it is deposited. Debris carried and deposited directly by a glacier is *till,* an unsorted mixture of particles ranging in size from clay to boulders.

Till accumulates in various topographic forms called *moraines.* Debris eroded from the walls of an alpine glacier is transported along the margin of the glacier as a *lateral moraine.* The joining of two lateral moraines at the confluence of a tributary glacier and the main glacier forms a *medial moraine* (fig. 3.18). Both lateral and medial moraines ultimately reach the glacier terminus or snout to form an *end moraine.*

Moraines are identified on photos of glaciers as dark bands in the wastage zone (fig. 3.17). Moraines are not visible in the accumulation zone because they are covered by the perennial snow that exists there. Lateral and medial moraines define the flow lines of an alpine glacier and cannot cross each other, but they may converge toward each other near the glacier terminus. An end moraine can remain long after the terminus where it was formed has retreated. Successive end moraines lying beyond the snout of a retreating glacier are called *recessional moraines.*

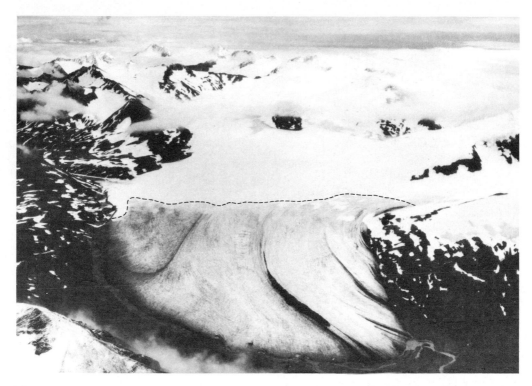

Figure 3.17   Oblique aerial photo of Snow Glacier, Kenai Peninsula, Alaska. Annual snowline shown as dashed line. (Photograph by Austin Post, U.S.G.S.)

Glaciers that terminate in a lake or the ocean produce *icebergs,* masses of glacier ice that become detached from the glacier terminus, a process called *calving.* Icebergs float freely but can become grounded when they are carried to shallow water by wind and currents. Icebergs eventually melt.

Landforms Produced by Alpine Glaciers

Erosional features usually dominate a terrain that was shaped by valley glaciers. A glacially eroded valley is *U-shaped* in cross-section, and its headward part may contain a *cirque,* an amphitheater-like feature. A cirque that contains a lake is called a *tarn. Hanging valleys* form where tributary glaciers once joined the trunk glacier, and waterfalls cascade from them to the main valley floor.

A narrow, rugged divide between two parallel glacial valleys is an *arête,* and the divide between the headward regions of oppositely sloping glacial valleys is a *serrate divide.* A pyramid-shaped mountain peak near the heads of valley glaciers or glaciated valleys is a *horn,* the most famous of which is the Matterhorn near Zermatt, Switzerland.

Moraines left by former alpine glaciers are rarely visible on standard topographic maps because the contour interval is usually larger than the height of most moraines.

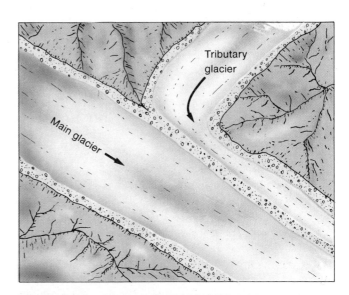

Figure 3.18   Lateral moraines join to become a medial moraine as a tributary glacier joins the main ice mass. (From Carla W. Montgomery, *Physical Geology,* 2d ed. Copyright © 1990 Wm. C. Brown Publishers, Dubuque, Iowa. All Rights Reserved. Reprinted by permission.)

Figure 3.19   Oblique aerial photographs of the South Cascade Glacier, Washington. (*A*) September 27, 1960; Neg. No. FR6025–50. (*B*) October 10, 1983; Neg. No. 83R1–188. (Courtesy of Andrew G. Fountain, U.S. Geological Survey, Tacoma, Washington.)

Exercise 15A. Mass Balance of an Alpine Glacier

Figure 3.19 shows two aerial photos of the South Cascade Glacier in the state of Washington. Photo *A* was taken on September 27, 1960, and photo *B* was taken on October 10, 1983. A comparison of the features shown in these two photographs reveals much about the recent history of this glacier. A longitudinal profile along the centerline of the glacier is shown in figure 3.20.

1. Draw a dashed pencil line on each of the photographs showing the boundary between the wastage and accumulation zones (ignore small patches of snow surrounded by bare ice). Label the line on photo *A* "annual snow line 1960," and the one on photo *B*, "annual snow line 1983."

2. During the time that has elapsed between the two dates of the photographs, has the snow line generally remained stationary, moved to a lower elevation, or moved to a higher elevation?
3. In comparing the two photographs, what is the evidence that the glacier has thinned in the wastage zone during the period 1960 to 1983?
4. Why are there no icebergs in the lake on photo *B*?
5. The annual net mass balances of the South Cascade Glacier for each of the years 1960 through 1982 are presented in graphic form in figure 3.21, and the actual values for annual net wastage or net accumulation, expressed in meters of water equivalent over the entire glacier surface, are given in table 3.2. Determine the algebraic sum of the net balance values in table 3.2, and use the result to refute or verify the evidence of glacial retreat based on the visual inspection of the two photographs.
6. Based on the data in table 3.2, in which year did the South Cascade Glacier come closet to being in equilibrium; that is, the net mass balance was almost zero?

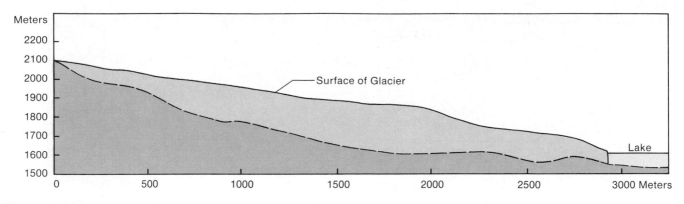

Figure 3.20 Longitudinal profile of the South Cascade Glacier during the 1965–66 budget year. No vertical exaggeration. (Based on U.S.G.S. Professional Paper 715–A, 1971.)

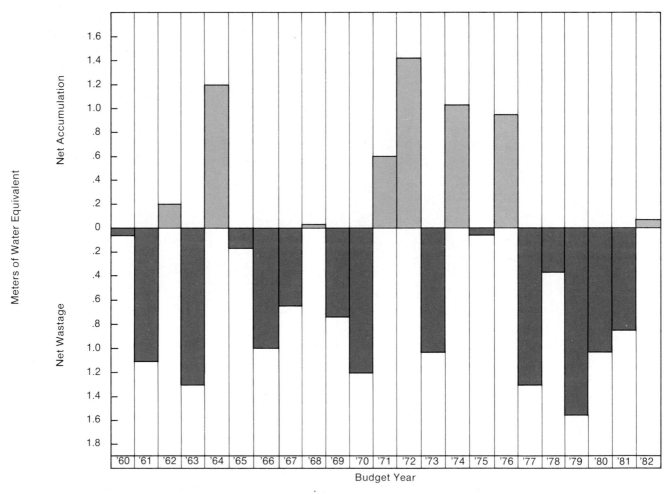

Figure 3.21 Annual net accumulation and annual net wastage on the South Cascade Glacier, Washington, for the budget years 1960 to 1982, based on data from table 3.2.

Table 3.2. Annual Net Mass Balances for the South Cascade Glacier, Washington, for the Budget Years 1960 to 1982.

BUDGET YEAR	NET BALANCE*	BUDGET YEAR	NET BALANCE*
1960	−0.50	1972	+1.47
1961	−1.10	1973	−1.03
1962	+0.20	1974	+1.02
1963	−1.30	1975	−0.05
1964	+1.20	1976	+0.95
1965	−0.17	1977	−1.31
1966	−0.98	1978	−0.38
1967	−0.63	1979	−1.56
1968	+0.02	1980	−1.02
1969	−0.73	1981	−0.84
1970	−1.20	1982	+0.08
1971	+0.59		

Source: Courtesy of Andrew G. Fountain, U.S. Geological Survey, Tacoma, Washington.

*Net balance values expressed as meters of water equivalent spread over the entire glacier.

Exercise 15B. Snowline, Moraines, and Glacier Flow

The Cordova Map (fig. 3.22) shows a number of alpine glaciers. The Heney Glacier flows from the lower left-hand margin of the map area in a northeasterly direction until it ends in a lake at the upper right-hand margin of the map area. A number of small tributary glaciers feed into the Heney Glacier, and a large, 3-mile-long, unnamed tributary joins the Heney in section 15 about 3.5 miles south of the northern boundary of the map area. Notice that the configuration of the glacier surfaces is depicted by blue contour lines that are continuations of the brown contour lines. The C.I. and all other principles used in the interpretation of contour lines are applicable to the blue contours. The pattern of brown stippling on parts of the glacier represents morainal material visible on the glacier surface.

1. Based on the relationship of medial and lateral moraines to the annual snow line on the Snow Glacier photograph (fig. 3.17), and assuming that the Cordova Map is based on glacier conditions at the *end* of the wastage season, determine the *maximum* elevation of the snow lines on the Heney Glacier and the McCune Glacier, and describe the reasoning used in arriving at your answer. (Assume that the snow line for each glacier is more or less coincident with a contour line across the glacier surface.)

2. Study the lower reaches of the Heney Glacier between the 1,700-foot glacier contour and the glacier terminus. Notice how the glacier contours from 800 to 1,200 feet outline a medial moraine. Trace the axis of this moraine by a red pencil line on figure 3.22 from the 1,600-foot contour to the glacier terminus. Show by red lines the two lateral moraines that formed the medial moraine.

3. Locate the *medial* moraine that lies between the 2,000- and 3,000-foot glacier contours on the west side of the Heney Glacier. Trace this moraine with a black pencil up-glacier to its logical point of origin. Show by black pencil the lower reaches of the two lateral moraines that formed it. Do the same for the four medial moraines above 1,500 feet on the tributary glacier that joins the Heney at about the 1,700-foot glacier contour.

4. Locate the point in the upper reaches of the Heney Glacier where the 4,500-foot contour line intersects the narrow rock outcrop south of the word *crevasses*. Assume that a large boulder becomes dislodged from the eastern edge of this rock island and falls onto the glacier surface. Using a blue pencil, draw a line on figure 3.22 from the point where this boulder begins its journey down-glacier to the point where it reaches the glacier terminus. (This line is a *flow* line and cannot cross a medial moraine.) If the average velocity of the glacier is 2 feet per day, and assuming that the glacier terminus remains fixed for the entire period, how long will it take for the glacier to transport the boulder along the route defined by your pencil line?

Continental Glaciation

Bodies of glacier ice covering most of Greenland and Antarctica are of continental proportions and are called *continental glaciers* or *ice sheets*. At times during the last one-and-a-half million years of the *Pleistocene Epoch* or *ice age,* continental glaciers covered much of North America, northern Europe, and parts of what is today the Soviet Union. The evidence of the past existence of continental glaciers lies in the deposits and landforms produced by them.

Depositional landforms of continental glaciers are recognized on topographic maps or stereopairs, but unlike the landforms produced by alpine glaciers, the full extent of which can be shown on a single topographic quadrangle, only a part of the landscape produced by a continental glacier can be covered on a standard topographic map. Nevertheless, contour maps with a scale of 1:24,000 to 1:62,500 and a contour interval of 25 to 50 ft adequately display the vestiges of continental glaciation.

One of the major objects in the study of continental glacial landforms is to reconstruct the configuration of various lobes of the Pleistocene ice sheets. A *glacier lobe* is a lobate appendage of the main ice sheet. These lobes range in breadth from less than 50 miles to more than 100 miles. By tracing end moraines over hundreds of miles, geologists can map the former extent of the ice margin of a single lobe as it existed during the waning phases of the ice age.

End moraines are belts of hummocky terrain a few miles in width that consist of till and intermixed deposits of stratified sand and gravel. While a glacial lobe occupied a position indicated by an end moraine, streams flowing from the glacier terminus deposited sand and

Figure 3.22   CORDOVA MAP
Part of the U.S.G.S. Cordova D–3
quadrangle, Alaska, 1953. Scale, 1:63,360;
contour interval, 100 feet.

Exercise 15C. Erosional Landforms Produced by Former Alpine Glaciers

The area shown in the Matterhorn Peak Map (fig. 3.23) was once occupied by a system of valley glaciers. The valleys produced by these glaciers now contain streams whose courses more or less coincide with the longitudinal axes of the former glaciers.

1. Trace the existing drainage lines in blue pencil. How do the valleys containing these streams differ in cross-section from the stream valleys south of the Mogollon Rim on the Promontory Butte Map of figure 3.6?

2. Sawtooth Ridge on the Matterhorn Map is the drainage divide between south-flowing streams such as Rock Creek and Spiller Creek, and streams flowing north to Robinson Creek.
   a) Visualize a topographic profile along the dashed line that defines the crest of Sawtooth Ridge. Does this profile lie more or less parallel to the contour lines or does it generally cut across them?
   b) Visualize a profile along the dashed line that approximates the divide on the Promontory Butte Map of figure 3.6. Does this profile more or less parallel the contour lines or does it generally cut across them?

c) Why do the two profiles on the two map areas differ?

3. Assign an appropriate physiographic name to each of the following features shown on the Matterhorn Map:
   a) Stanton Peak (SE quadrant of the map).
   b) Spiller Lake (SW quadrant of the map).
   c) The part of Horse Creek Valley lying up-valley from the waterfalls at about the 7,600-foot contour line (NE quadrant of the map).
   d) The Cleaver (central part of the map).
   e) Sawtooth Ridge.

4. The topography of the Matterhorn Peak Map area is dominated by landforms produced by glacial erosion. During the time when the former valley glaciers were receding, it is not unreasonable to assume that recessional moraines might have been formed. Remnants of an end moraine are represented by the 7,120-foot contour line north of the Twin Lakes Campground in the extreme northeast corner of the map. With this one exception, no other recessional moraines are in evidence on the entire map. Suggest a reason for this.

gravel beyond the front of the glacier as an *outwash plain.* Closed depressions lying in end moraines and on outwash plains are sites of former stagnant ice masses that became separated from the retreating glacier. When these ice masses melted, they left depressions in their place called *ice block pits.* An outwash plain with many pits, some of which contain lakes, is called a *pitted outwash plain.* An ice block pit in an end moraine is called a *kettle.*

A stream flowing in a stagnant ice channel deposits sand and gravel along the ice walls of the channel. When the ice melts, the stratified deposits remain in a variety of topographic forms called *ice-contact deposits.* One form of ice-contact deposit is a *kame,* a knob, hummock, or conical hill composed of sand and gravel. A chaotic assemblage of kames and kettles common to many end moraines is called *knob and kettle topography.*

Other products of continental glaciation are eskers and drumlins. An *esker* is an ice-contact deposit in the form of a ridge of sand and gravel believed to have been formed beneath the glacier surface by a sediment-laden stream flowing through an ice tunnel. A *drumlin* is an elongate and streamlined hill of till whose long axis is parallel to the direction of ice movement. The steep side of one of the elongate ends of a drumlin lies toward the direction from which the ice was flowing, and the more gentle side of the other end of the drumlin axis lies in the downstream direction of ice flow. Drumlins are molded at the base of the ice sheet and commonly occur in swarms called *drumlin fields.*

Figure 3.23 MATTERHORN PEAK MAP
Part of the U.S.G.S. Matterhorn Peak
quadrangle, California, 1956. Scale,
1:62,500; Contour interval, 80 feet.

Exercise 16A. Moraines and Outwash Plains

The topography shown on the Whitewater Map (Fig. 3.26) was produced by the last advance and retreat of the continental ice sheet. Three general terrain types are present: (1) ground moraine, (2) end moraine, and (3) pitted outwash plain. Because the area covered by the map is so small compared to the total area covered by the continental glacier, the relationship of the Kettle moraine (an end moraine) to the position of the ice front at the time the moraine was built is not immediately clear from a casual inspection of the map.

Let us assume that the Kettle moraine on the Whitewater Map is part of an end moraine that was deposited by the "Kettle lobe" as shown in the sketch map of figure 3.24. On that sketch map, two possible locations of the Whitewater Map are shown as A and B. Each shows the Kettle moraine trending across the map area from southwest to northeast, as it does on the Whitewater Map. Beyond the end moraine of figure 3.24, outwash associated with the "Kettle lobe" is shown. In addition, we can assume that the area formerly underlain by the "Kettle lobe" will be covered with ground moraine.

1. Study the Whitewater Map and the topographic profile of figure 3.25 and determine which of the two locations, A or B on figure 3.24, is the one that most likely represents the Whitewater Map. State in concise terms the reasons for your choice.

2. Draw the ice-surface *profile* of the "Kettle lobe" on figure 3.25 as it may have been when the "Kettle lobe" was building the Kettle moraine. (In drawing the ice-surface profile of the "Kettle lobe," assume that the ice slopes upward from the highest point on the end moraine until it attains a thickness of about 500 feet in a horizontal distance of about 4 miles.)

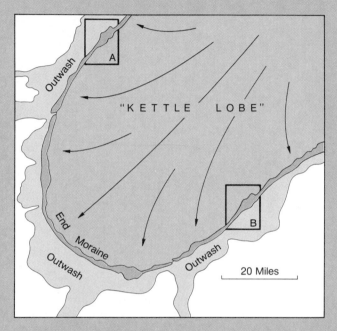

Figure 3.24 Map of the hypothetical "Kettle lobe" showing two possible locations of the Whitewater Map (fig. 3.26), A or B. (See Exercise 16A for instructions.)

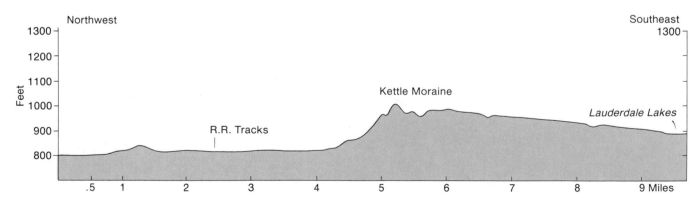

Figure 3.25 Generalized topographic profile from northwest to southeast across the Kettle moraine of the Whitewater Map (fig. 3.26). (Vertical scale exaggerated.)

Figure 3.26 WHITEWATER MAP
Part of the U.S.G.S. Whitewater quadrangle, Wisconsin, 1960. Scale, 1:62,500; contour interval, 20 feet.

Exercise 16B. Drumlins

Figure 3.27 is a stereopair of part of a drumlin field in New York state. The upper photo in the stereopair covers the area of the Sodus Map of figure 3.28. Study the stereopair with your stereoscope in conjunction with the topographic map. (The steepness of slopes on the stereopair is exaggerated in stereovision.)

1. What was the general direction of flow of the continental glacier across the map area? What is the evidence to support your answer?

2. Locate the main road that roughly parallels the western boundary of the map, Norris Road, and the north-south railroad tracks that pass through Zurich. On the upper photo of the stereopair, show these roads in red pencil and the railroad in black pencil. Use sharp pencils so as not to obscure your stereovision.

3. To what extent has the topography influenced the routes of the two roads and the railroad?

4. To what extent has stream erosion modified the topography of the area?

5. What geologic material is likely to predominate in the areas occupied by the drumlins?

6. A pair of power lines extends in an east-west direction near the southern margin of the map.
   a) How is the route of the power lines identified on the stereopair?
   b) Is the map route of the power lines controlled by the topography?

7. Draw the outline of Mud Pond on the upper photo. What has happened to Mud Pond during the 34 years that have elapsed between 1952, when the map was published, and 1986, when the photos were taken?

8. What is the R. F. of the stereopair?

Figure 3.27 Stereopair of part of a drumlin field in New York State. The upper photo covers the same area as the topographic map in figure 3.28. (Photos taken April 14, 1986. HAP 85, series 437608, photos 212-159, 160.)

Figure 3.28 SODUS MAP
Part of the U.S.G.S. Sodus quadrangle, New York, 1952. Scale, 1:24,000; contour interval, 10 feet.

Exercise 16C. Ice-Contact Deposits

Figure 3.29 is a stereopair of an esker in Michigan.

1. Trace the crest of the esker in red pencil.

2. Visualize the esker crest in profile. Use a black pencil to draw a single closed contour line showing the highest part of the esker crest. Draw the contour line on the right-hand photo while viewing the stereopair in stereovision.

3. Just east of the bend in the esker is a conical hill or knob. What is the name for this ice-contact deposit?

4. What constructional materials might be available from the esker and the ice-contact deposit?

Figure 3.30 shows two north-south trending eskers.

5. Trace the crest of each esker with a light-colored, felt-tip highlighting pen.

6. What local name is used for the eskers?

7. What map evidence verifies the assumption that eskers are ice-contact deposits?

8. The crest of the esker east of the Penobscot River does not lie at a constant elevation along its course. Given this observation, why is it necessary to invoke the presence of glacier ice to account for the origin of an esker?

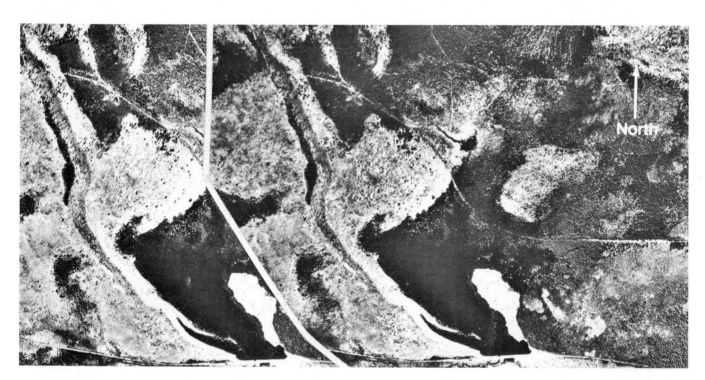

Figure 3.29 Stereopair of part of an esker in Michigan. Scale, 1:24,000. (Courtesy of U.S. Geological Survey.)

Figure 3.30 PASSADUMKEAG MAP
Part of the U.S.G.S. Passadumkeag
quadrangle, Maine, 1960. Scale, 1:62,500;
contour interval, 20 feet.

Landforms Produced by Wind Action

Wind is a geologic agent of erosion, transportation, and deposition. This section deals with landforms produced by wind erosion and wind deposition that are visible on standard topographic maps and aerial photographs or other images.

Blowouts

A *blowout* is a shallow depression caused by the removal of sand or smaller particles through the process of *deflation*. Deflation is the removal of loose surface material by wind that has a velocity sufficient to carry sand or silt particles aloft and transport them to another location. Deflation cannot lower the ground surface below the water table.

Blowouts are irregular in map view, are formed where there is little or no vegetation, and commonly are defined by closed contour lines on a topographic map. If a blowout contains a shallow pond or lake, it can be inferred that the water table has risen after the blowout formed. A change to a more humid climate (i.e., increased rainfall) could cause the water table to rise.

Sand Dunes

A *sand dune* is a ridge or mound of sand deposited by wind. Winds that blow consistently from the same direction are called *prevailing winds*. Sand dunes formed by prevailing winds have a distinctive topographic profile. The side of a dune ridge facing the wind is called the *windward* side, and is characteristically less steep than the slope on the *downwind* or *leeward* side. The leeward slope is referred to as the *slip face,* and its angle with the horizontal is about 30 degrees. The identification of the slip face of a dune allows one to infer the direction of the prevailing winds that formed it. Dunes that do not exhibit a slip face are formed by winds that blow in different directions at different times but with comparable speeds.

Dunes occur mainly in arid or semiarid regions with little or no vegetational cover. Dunes also occur along coastal regions where sand beaches provide a source of sand for the prevailing onshore winds. Dunes may occur in single isolated geometric forms or they may exist in swarms called *dune fields*. The factors that control the geometric form of a dune are wind direction or directions and velocity, supply of sand, and the extent of vegetational cover.

Types of Sand Dunes

Sand dunes formed by prevailing winds occur in a variety of geometric shapes. Figure 3.31 shows three common forms that can be identified on maps or photos. A *barchan* is crescent shaped in map view, and the horns of the crescent point downwind. A *transverse dune* is a linear ridge with its long dimension oriented at right angles to the prevailing wind direction. The slip face of a transverse dune provides the basis for determining the wind direction unambiguously. A *longitudinal dune* has a linear form with its long axis parallel to the wind direction.

Barchans are common on flat desert surfaces with a sparse supply of sand and little or no vegetation. Transverse dunes occur where the sand supply is abundant and vegetation is sparse. Longitudinal dunes occur in deserts with strong winds that vary only slightly from a single direction. Some elongated sand ridges in coastal regions appear to be longitudinal dunes, but on close inspection they are seen to be ridges of windblown sand anchored by vegetation and separated by wind-scoured troughs parallel to the dunes.

Active dunes are those that migrate in a downwind direction or are constantly changing in shape in response to multiple wind directions at different times. *Inactive dunes* are those that have become stabilized by the growth of a vegetational cover to the extent that their migration ceases. Stabilized dunes are indicative of a climatic change to more humid conditions. When patches of vegetation on inactive dune fields die from drought or other causes, the exposed sand is deflated to form local blowouts.

Coastal dunes that invade an area occupied by stream valleys sloping toward the coast may block the valleys and cause ponding of the stream into small lakes or ponds. Such water-filled depressions are not to be confused with blowouts.

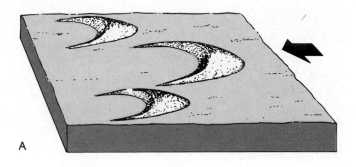

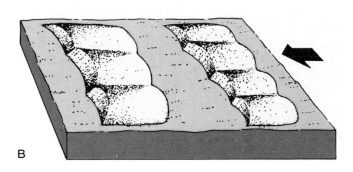

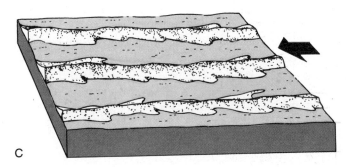

Figure 3.31 Types of sand dunes. (*A*) Barchans. (*B*) Transverse dunes. (*C*) Longitudinal dunes. (From Charles C. Plummer and David McGeary, *Physical Geology,* 4th ed. Copyright © 1988 Wm. C. Brown Publishers, Dubuque, Iowa. All Rights Reserved. Reprinted by permission.)

Exercise 17A. Barchans

The crescent-shaped dunes in figure 3.33 are active barchans on the desert floor west of the Salton Sea in southern California. Ground measurements show that they moved between 325 and 925 feet during a 7-year period. Study the stereopair of figure 3.33 and compare it with the topographic map of figure 3.32. Notice the difference in scales between the stereopair and the topographic map.

1. What is the prevailing wind in the area? How is it determined?

2. Assuming that an airstrip should be aligned approximately parallel to the prevailing wind direction, is the airstrip properly aligned?

3. The stereopair reveals many more sand dunes than the number visible on the topographic map. Account for this difference.

4. Assuming that the dunes will continue their direction and rate of migration within the limits already stated, what will be the shortest and longest times, in years, for the barchan about 1/2 mile from the west end of the landing strip to reach the landing strip?

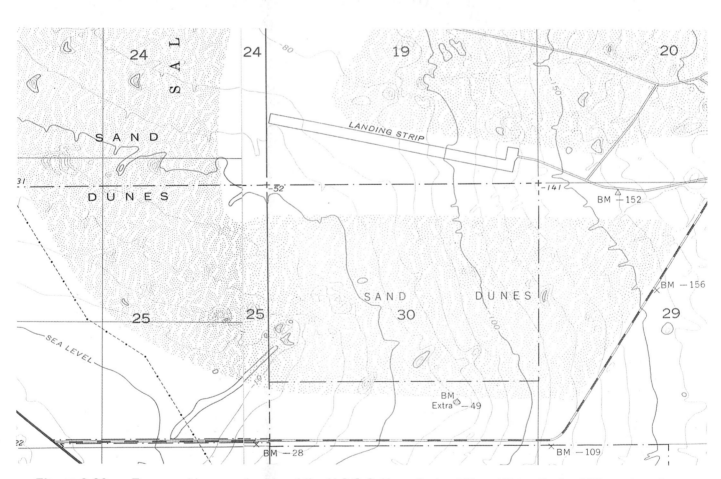

Figure 3.32   Topographic map of parts of the U.S.G.S. Kane Spring NE and Kane Spring NW quadrangles, California, 1956. Scale, 1:24,000; contour interval, 10 feet.

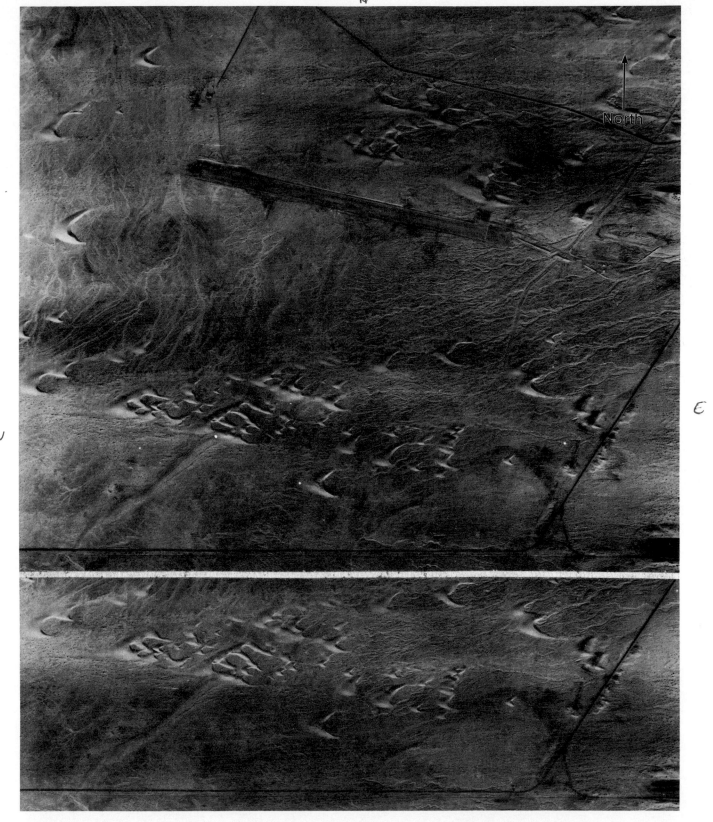

Figure 3.33   Stereopair of Kane Spring sand dunes, California. Scale, 1:20,000; taken, November 10, 1959.

Exercise 17B. Coastal Dunes

Figure 3.34 is a topographic map of coastal dunes in southern California north of Santa Barbara. Figure 3.35 is a stereopair of the southern two-thirds of the map area. Study the stereopair under stereovision and identify features on it that are labeled on the map. When you first view the stereopair under stereovision, you may see the sand dunes in *reverse topography*. In reverse topography the hills appear as depressions, and the steep slopes appear to be facing in the direction opposite to their true orientation. In the stereopair of figure 3.35, the steep slopes on the sand dunes face *away* from the coast in true relief. You may have to "coax your mind" to bring the relief into true perspective. By turning the stereopair upside-down, you may find it easier to bring the true relief into view.

1. What kind of dunes are those with their ridges lying parallel to the coast?

2. The narrow linear ridges next to the coast have their axes aligned at an angle of about 70 degrees to the coastline. Dark lines of vegetation are associated with the ridges. Do the vegetational lines occur generally on the top of the ridges or between them? What is the likely origin of the ridges? Are they longitudinal dunes?

3. Draw an arrow in black pencil on figure 3.34 showing the inferred direction of prevailing winds.

4. What is the source of the sand contained in the dunes?

5. What is the likely origin of Little Oso Flaco Lake and Mud Lake?

Figure 3.35   Stereopair of sand dunes on the California Coast north of Santa Barbara. Scale, 1:28,800. (Photos taken June 27, 1963. Series GS-VASJ, numbers 1–262 and 1–265.)

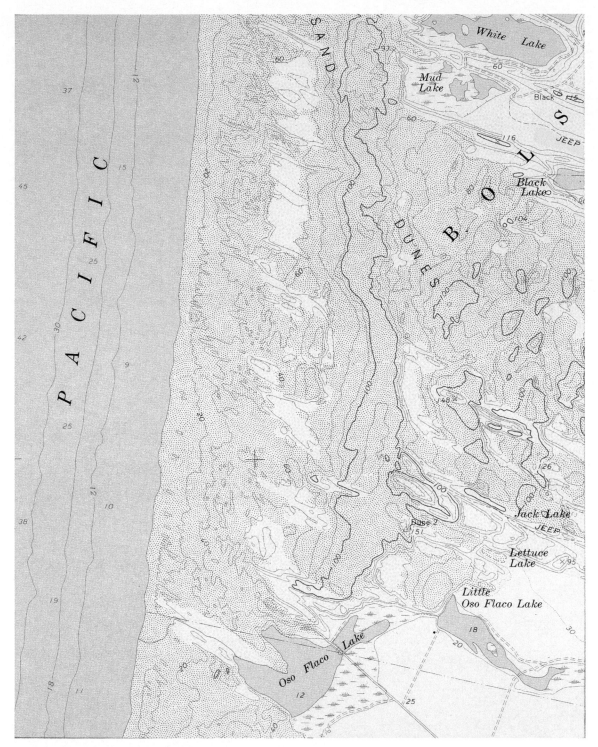

Figure 3.34 OCEANO MAP
Part of the Oceano quadrangle, California, 1965. Scale, 1:24,000; contour interval, 20 feet.

Exercise 17C. Inactive Dune Fields

Figure 3.36 is a Landsat image, similar to a black-and-white aerial photograph, taken from an altitude of 570 miles over western Nebraska in the winter of 1973. This area is known as the Sand Hills because the topography consists of a massive dune field that is now virtually inactive because of a lush vegetational cover of prairie grasses. The dunes were formed during a time when the rainfall was less than at present and unable to support the grasses that grow there now. A heavy cover of snow enhances the topography.

The relief of the Landsat may appear inverted. If the dune ridges appear to be depressions, turn the photo upside-down so that the north arrow points toward you and the true relief will "pop" into view.

The rectangular area in the northwestern corner of the map is the area covered by the Ashby Map of figure 3.12. The two dark bands in the lower part of the image are the North and South Platte Rivers which converge toward the town of North Platte, Nebraska, near the eastern margin of the image.

1. What is the R. F. of the Landsat image?

2. What are the only features visible on both the Landsat image and the Ashby Map?

3. Some of the interdune depressions visible on the Ashby Map contain lakes. Assuming that some of the interdune depressions were sites of deflation during the period of dune formation, what does the presence of the lakes imply about the elevation of the water table when the dunes were active?

4. The topography of the dunes on the Ashby Map reveals that many of them are irregular in map view and contain no clear-cut distinction between gentle and steep slopes. However, those that are elongated in a general east-west direction are typical of those visible on the Landsat image. Study the contour lines on the north and south side of the east-west trending dunes on the Ashby Map, determine which slopes are the steepest, and infer the direction of the prevailing winds at the time of deposition.

5. Based on the answer to question 4, draw arrows about 1/4 inch long at several points on the Landsat image to show the direction of the prevailing winds when the dune field was formed.

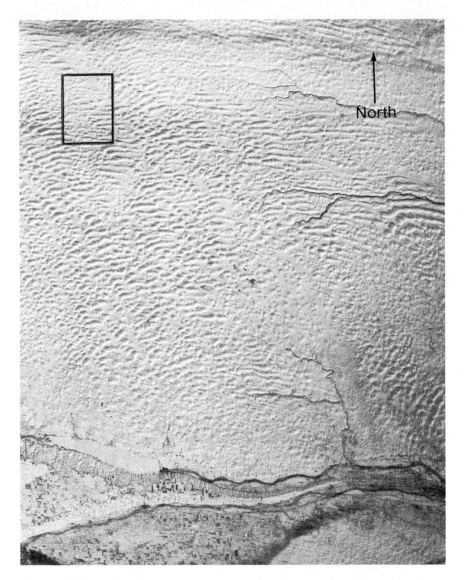

Figure 3.36 Landsat image of part of western Nebraska made on January 9, 1973 from ERTS-1 at an altitude of 570 miles. The rectangle in the northwest corner is the area covered by the Ashby Map of figure 3.12. (NASA ERTS E−1170−17020, Courtesy of NASA and the U.S.G.S. EROS Data Center, Sioux Falls, South Dakota 57198.)

Modern and Ancient Shorelines

The interaction between the ocean or a large lake and the land occurs at the coastline or shoreline. Waves generated by winds eventually reach the shore, where the wave energy is dissipated. Waves breaking along a shore are geologic agents of erosion, transportation, and deposition. This section deals with the modern and ancient landforms produced by one or a combination of these agents as they are portrayed on topographic maps and aerial photographs.

Depositional Landforms Produced by Wave Action

A wave crest moving toward shore is bent or *refracted* as it approaches shallow water (fig. 3.37). As the successive incoming waves break just off shore, a *longshore current* is initiated (fig. 3.38). A wave breaking on the shore carries sand particles up the beach in the direction of wave movement. When the water from the spent wave flows back down the beach, it follows a course controlled by the slope of the beach. Sand particles exposed to this alternate movement are moved along the beach, a process known as *beach drift* (fig. 3.38). The longshore current moves fine sand and beach drift moves coarser sand particles parallel to the shoreline. When the longshore current slows due to deeper water associated with an indentation on the coast such as a bay or estuary, sand is deposited in the form of a *spit*. A spit may grow across the mouth of a bay to form a *baymouth bar* (fig. 3.39). A spit that is curved shoreward is a *recurved spit*. Other sandy features along the coastline include *beaches* and *barrier islands*. Indentations along the coastline that become isolated from the main body of water become lakes or lagoons. *Lagoons* are shallow water bodies lying between the main shoreline and a barrier island. A *barrier island* is an elongate sand island parallel to the shoreline. A break or passageway through a bar or barrier island is a *tidal inlet*, so called because it allows currents to flow into the lagoon during rising tides and out of the lagoon during falling tides. Lagoons and coastal lakes eventually become filled with sediment and are transformed into *mud flats, marshes,* and *wetlands.*

Beach sand is attacked by onshore winds that carry the sand inland to form coastal dunes. The sand lost to wind action is replenished by wave action and beach drift.

Erosional Landforms Produced by Wave Action

An initial shoreline consists of bays and headlands. A *headland* is a part of the coast that juts out into the lake or ocean. Wave refraction concentrates the energy of waves against headlands (fig. 3.37). The landforms produced by

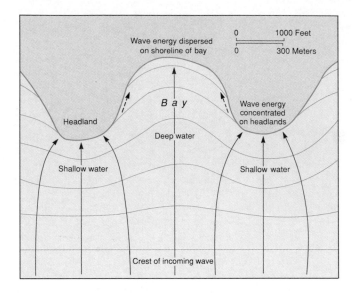

Figure 3.37 Schematic map showing the refraction of waves approaching an irregular shoreline. Wave energy is concentrated on the headlands and dispersed in the bays. The arrows represent lines of equal wave energy. These are equally spaced where the water depth is below the wave base but curved toward the headlands when one part of the wave crest strikes shallow water before the rest of the wave. Dashed arrows show the direction of beach drift.

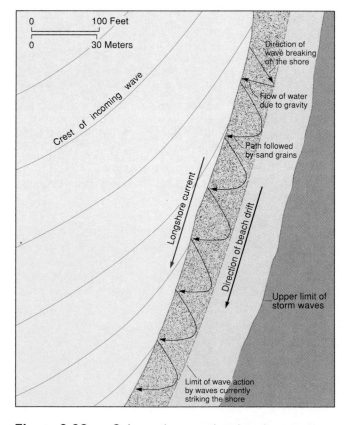

Figure 3.38 Schematic map showing the relationship of refracted waves breaking on the shore to the longshore current and beach drift.

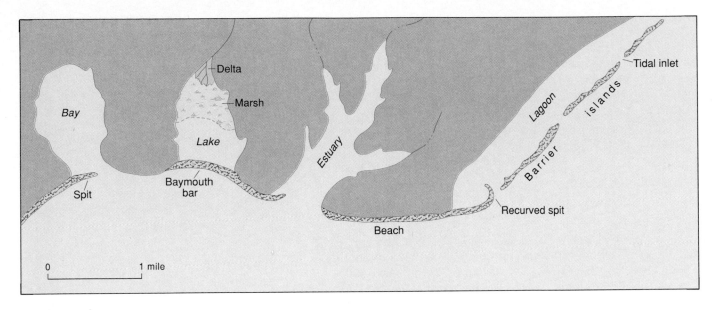

Figure 3.39   Schematic map showing landforms associated with wave action.

this intensified wave action are *wave-cut cliffs,* seaward-facing escarpments formed by wave erosion at their bases, and *stacks,* rocky pillars that are remnants of retreating wave-cut cliffs.

As a wave-cut cliff retreats under constant wave erosion at its base, a wave-cut platform is formed. A *wave-cut platform* is a gently sloping rock surface lying below sea level and extending seaward from the base of a wave-cut cliff.

Evolution of a Shoreline

Sea level was some 200–300 feet lower than at present during parts of the Pleistocene when continental glaciers covered about 30 percent of the earth's land surface. As the Pleistocene ice sheets melted, sea level rose to its present level. A change in sea level because of an increase or decrease in the volume of water in the oceans is called a *eustatic rise* or a *eustatic fall* in sea level.

The shorelines around many coasts at the end of the Pleistocene were irregular and consisted of long embayments formed by the encroachment of the sea into the mouths of rivers. These *drowned river mouths* were separated by headlands, which were the sites of intense wave erosion. So long as sea level remained unchanged, the headlands became wave-cut cliffs, wave-cut platforms were formed, and drowned river mouths were cut off from the sea by the growth of spits and baymouth bars. Through the process of wave erosion and deposition, the original post-Pleistocene shoreline changed to a straighter form.

Sea level does not remain constant over time, however. A local uplifting of the land along a coast has the same effect as a drop in sea level. Because it cannot always be determined whether sea level has risen or fallen eustatically, whether the land has risen or subsided, or

whether a combination of these events has occurred, it is common practice to refer to *relative* changes in sea level.

Generally, a relative fall in sea level produces an *emergent shoreline,* and a relative rise in sea level produces a *submergent shoreline.* The shorelines of the world's coasts at the end of the Pleistocene were generally submergent, but in coastal regions that were formerly near or within the borders of the continental glaciers, the land began to rise in response to the retreat of the ice caps whose weight had depressed the earth's crust. This uplift did not begin immediately upon retreat of the ice, so there was time for erosional and depositional shore features to be built. These shoreline features now lie above modern sea level, where they are exposed to the normal processes of subaerial weathering and erosion which modify and eventually destroy them.

Deltas

A *delta* is a nearly flat plain of riverborne sediment extending from the river mouth to some distance seaward. The deltaic sediments are deposited by *distributaries,* channels that branch out from the main channel of the river. Distributaries are bordered by *natural levees,* narrow ridges of river sediment built by a river overflowing its banks during flood stages. A delta with many distributaries is called a *birdfoot delta* because of its resemblance in plain view to the outstretched claws of a bird's foot.

New distributaries are formed from time to time, and older distributaries are abandoned. An entire delta may be abandoned when the course of the river feeding it shifts to a new course many miles upstream. The abandoned delta then becomes an *inactive delta.* The inactive delta no longer receives sediment and is modified by wave action

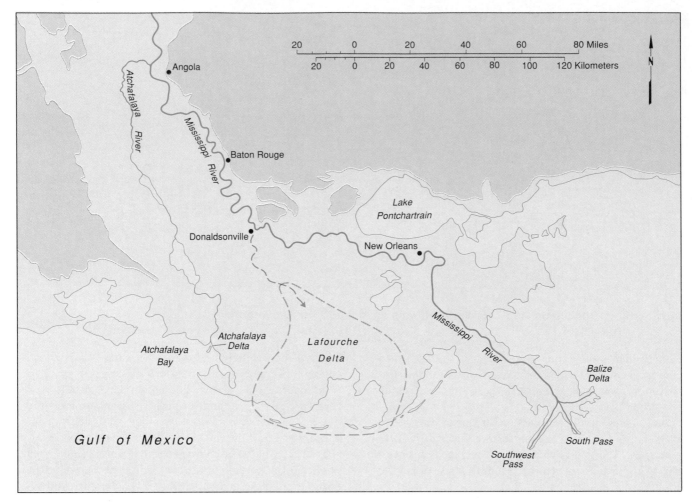

Figure 3.40   Map of the Mississippi River valley and deltas associated with it. (From W. D. Thornbury, *Regional Geomorphology of the United States.* Copyright © 1956 John Wiley & Sons, New York.)

to the extent that its seaward margin becomes smooth and it loses its characteristic birdfoot form.

An active delta expands seaward if the supply of sediment is too large to be dissipated by wave action and longshore currents. A shoreline that moves seaward by deltaic growth is called a *prograding shoreline.* Progradation ceases when a delta becomes inactive, and its seaward margin may in fact retreat due to wave action and subsidence of the deltaic deposits. Eventually, waves and currents redistribute sand along the old delta front to create lagoons behind barrier islands, tidal inlets, and saltwater marshes. An inactive delta can be reactivated if the river course shifts upstream and discharges once again into the site of the former active delta.

The Mississippi Delta

The Mississippi River drainage system carries an enormous load of sediment that eventually is dumped into the Gulf of Mexico along the coast of Louisiana. The delta currently being built by the Mississippi River is the Balize or Birdfoot delta, and it has been functional during the last 800–1,000 years. Over the last 7,500 years, four previous deltas, now inactive, have been built by ancestral courses of the Mississippi River. The most recent of these inactive deltas, the Lafourche delta, was active between 2,500 and 800 years ago. It lies adjacent to and on the west side of the Balize delta (fig. 3.40).

In the mid-twentieth century, the Mississippi River was discharging about 25 percent of its flow through the Atchafalaya River, some 100 miles north of Atchafalaya Bay (fig. 3.40). This flow would have increased until the entire volume of the Mississippi River would have discharged into the Gulf of Mexico through the Atchafalaya River. Had this natural event occurred, the modern course of the river south of Angola and the Balize delta would have been abandoned. However, this natural diversion was forestalled in 1963 when the U.S. Army Corps of Engineers completed a control dam at the diversion site. In spite of this human intervention, however, the sediment deposited by the Atchafalaya River into Atchafalaya Bay has started what might become the next delta of the Mississippi River.

Ancient Shorelines Around the Great Lakes

During the retreat of the continental glaciers from the Great Lakes region, precursors of the modern Great Lakes formed around the lobate front of the ice margin (fig. 3.41). The shorelines of these ancestral Great Lakes are preserved in beach ridges and wave-cut cliffs that now lie above the elevations of the present Great Lakes. Several different ancestral lakes were formed at various times during the retreat of the ice, and each of these lake stages has been assigned a name to distinguish one from the other. Figure 3.41 shows the extent of glacial lakes Maumee and Chicago in the basins now occupied by Lake Erie and Lake Michigan. The levels of these and other stages can be determined by using the elevations of their shorelines to infer the elevations of the water planes of the lakes. For example, the base of an abandoned wave-cut cliff would mark the water plane of the lake that produced it.

Where two or more ancestral shorelines are related to the basin of a modern lake, we will assume that the highest shoreline is the oldest and the lowest is the youngest. While this relationship is not universally true for all the former lake stages, it will suit our purposes for this discussion. Note that a higher lake may have a different outlet; for instance, Lake Maumee and Lake Chicago and the other drainage systems presented on figure 3.41 had outlets to the Mississippi River at this time, because the present outlet of the lakes through the St. Lawrence River was blocked by the Ontario lobe.

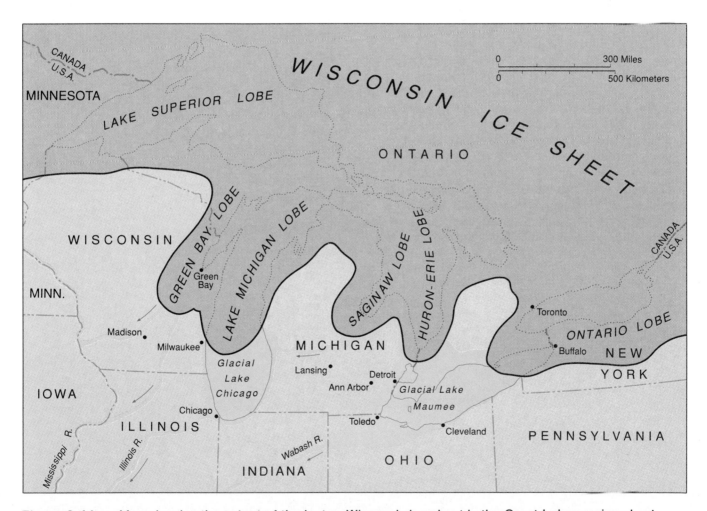

Figure 3.41   Map showing the extent of the last or Wisconsin ice sheet in the Great Lakes region about 14,000 years ago. The glacial lakes Chicago and Maumee drained through outlets to the Gulf of Mexico via the Mississippi River. (From Zumberge and Nelson, 1972, *Elements of Geology*, 3d ed. Copyright © John Wiley & Sons, New York.)

The Modern Great Lakes

The Great Lakes straddle the U.S.-Canadian border and have a combined surface area of almost 100,000 square miles (fig. 3.42). The lakes came into being during the retreat of the Pleistocene ice sheet that once covered the entire area more than 10,000 years ago. The volume of water in the modern lakes is mainly a function of surface runoff from the surrounding drainage basin, evaporation from the lakes themselves, and the outflow of water to the Atlantic Ocean via the St. Lawrence River. The volume of water stored in the Great Lakes is reflected in their water levels. These levels are published monthly for each lake by the U.S. Army Corps of Engineers (fig. 3.48) and are a matter of public record.

The shorelines of the Great Lakes are sites of thousands of vacation homes and permanent dwellings. Those with lake frontage are desirable from an aesthetic point of view, but they are in considerable jeopardy during periods of high water levels when storm waves destroy sandy beaches formed during low water stages, and cause severe recession of the bluffs and cliffs by wave erosion at their bases.

Much of the eastern shoreline of Lake Michigan, for example, is characterized by steep bluffs that are formed in old sand dunes and other unindurated Pleistocene sediments. Some of the bluffs rise to more than 100 feet above the lake. Private dwellings on the tops of these bluffs were built during periods of low water when the shores between the bases of the cliffs and the water's edge were characterized by wide sandy beaches. Those who purchased or constructed homes during low water stages believed that the wide beaches fronting their properties were a permanent part of the landscape and provided adequate protection from any future wave erosion. These were false assumptions, as many who owned property on the shores of Lake Michigan learned at great cost during the period 1950 through 1987.

Three times during this period, 1952–1953, 1972–1976, and 1985–1986, the water levels of Lake Michigan stood at extraordinarily high levels, and twice during the same period, 1958–1959 and 1964, the levels were extremely low (fig. 3.48). The water level of 581.6 feet in 1986 was the highest on record since 1900, and the water level of 575.7 feet in 1964 was the lowest on record for the same period. Thus, in the 22-year period between 1964 and 1986, the level of the lake varied by about 6 feet, mainly by natural causes.

The damage to property during high water stages is all too apparent in the photograph of figure 3.49. The house in figure 3.49 was abandoned by the time it was photographed in 1986.

One might ask why these houses and hundreds of others like them were built in the first place. The answer lies in the lack of understanding of the relatively short time it takes for the lake level to change drastically, and the inability of anyone to predict future levels over a time period of a decade or so. No one would think of building a house next to the one in figure 3.49 today. The consequences of such folly are all too apparent. But, when the houses along Lake Michigan and other Great Lake shores were built during low water stages, they seemed secure from the damage and destruction to which they were subjected in later years.

The lesson to be learned from this is that those who contemplate purchasing or building homes should be aware of geologic hazards, and endeavor to acquire as much information as possible about building sites from the public record before proceeding.

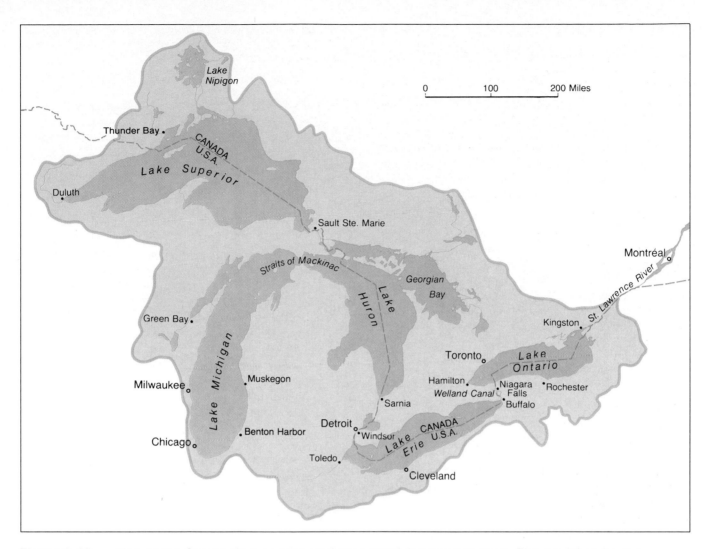

Figure 3.42   Map of the Great Lakes and the land area supplying runoff to them. The five lakes have a combined surface area of almost 100,000 square miles, and the drainage basin containing them is roughly twice that size.

Exercise 18A. Coastal Processes and Shoreline Evolution

The aerial photograph of figure 3.35 shows three waves approaching the shore.

1. Draw a red line along the crest of each of the three waves, and show by a red arrow on the water surface the direction of the longshore current. Show by a black arrow on the beach the direction of beach drift.

2. Were the waves on figure 3.35 produced by a wind blowing in the same direction as the prevailing winds that formed the sand dunes?

3. The Point Reyes Map of figure 3.43 shows part of the California coast near San Francisco. North is toward the bound margin of the map. The configuration of the contour lines along Point Reyes Beach reflects the elongate orientation of coastal sand dunes. Draw several short red arrows along the coast that show the prevailing wind direction that formed the dunes.

4. Assume that wind blowing in the direction indicated by the red arrows you have drawn produces waves that strike the shoreline of Point Reyes Beach. Based on the wave refraction pattern of these hypothetical waves, draw a red arrow along the beach showing the direction of beach drift.

5. Draw a red arrow along Limantour Spit to show the direction of beach drift and a dashed red arrow to show the direction of the longshore current just seaward of the spit.

6. What is the origin of the horseshoe-shaped lake east of the D Ranch?

7. Explain the irregular shape of the embayment formed by Drakes Estero and its many tributaries.

8. Use black dashed lines to extend seaward the 80-foot land contours along the west shore of Drakes Bay between the entrance of Drakes Estero and Point Reyes to show how the shoreline might have appeared before wave action produced the present smoothly curved shore. What is the name applied to the features outlined by the extension of the 80-foot contours?

9. What features in Drakes Estero indicate the ultimate fate of Drakes Estero and its many arms?

10. Point Reyes is marked by a steep cliff on its seaward side. What is the origin of this cliff and the many small islands lying just off shore at its base?

Figure 3.43 POINT REYES MAP
Part of the U.S.G.S. Point Reyes quadrangle, California, 1954. Scale, 1:62,500; contour interval, 80 feet.

Exercise 18B. Deltas of the Mississippi River

Figure 3.44 shows the Balize delta of the Mississippi and the plume of sediment being discharged to the Gulf of Mexico. The major distributaries of the Balize delta are called *passes,* and their names and the names of associated interdistributary bays are shown in figure 3.45. The inactive Lafourche delta lies on the western flank of the Balize delta.

1. What are the narrow ridges visible on either side of Southwest Pass in figure 3.44?

2. The light blue color around the Balize delta in figure 3.44 is a plume of suspended sediment. What is the immediate resting place of this sediment?

3. If the Balize delta should be abandoned in favor of the Atchafalaya delta, describe the changes that would occur on the seaward edge of the Balize delta.

4. The Mississippi River is a major artery of commerce connecting the Gulf of Mexico with New Orleans and Baton Rouge. Describe the impact on the river between the delta and the cities along its course if human intervention had not been imposed on the site of natural diversion of the Mississippi River near Angola.

5. What features of the Lafourche delta indicate that it was once an active delta?

Figure 3.44 False color image of the Mississippi Delta made from Landsat on April 3, 1976. (Courtesy of NASA and the U.S.G.S. EROS Data Center, Sioux Falls, South Dakota 57198.)

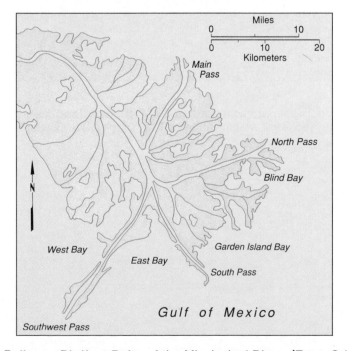

Figure 3.45 Map of the Balize or Birdfoot Delta of the Mississippi River. (From Orin H. Pilkey, et al., *Coastal Land Loss Short Course in Geology,* Vol. 2. Copyright © 1989 American Geophysical Union, Washington, DC.)

Exercise 18C. Ancestral Lakes of Lake Erie

The North Olmstead Map (fig. 3.47) covers an area just west of Cleveland, Ohio. Detroit Road and Center Ridge Road lie along two shorelines ancestral to Lake Erie. These will be referred to as the Detroit shoreline and the Center Ridge shorelines. The Center Ridge shoreline was formed about 13,000 years ago during the waning phases of the Pleistocene, approximately 1,000 years after glacial lake Maumee.

Figure 3.46 is a north-south profile showing the location of Center Ridge Road and Detroit Road. The profile is aligned along Forest View Road just west of Forest View School, and extends north to the 30-ft depth contour in Lake Erie and south to the 740-ft contour line.

1. Draw the line of profile with a black pencil on figure 3.47.

2. With reference to the profile, compare the Detroit and Center Ridge shorelines with the shoreline of Lake Erie. Which of these three shorelines are erosional and which are depositional?

3. Use a sharp pencil and a straightedge to draw the water surfaces of the lakes that produced the two ancestral shorelines. Write the elevations of each above the two lines you have drawn.

4. Draw a solid black line on the profile between Detroit Road and Wolf Road to show what the original profile might have looked like before the Detroit shoreline was formed.

5. Based on the elevations of the two shorelines, which is older?

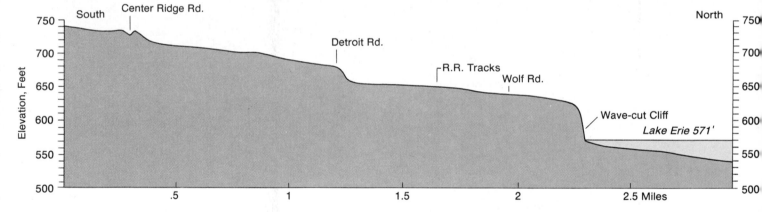

Figure 3.46 North-south topographic profile based on the North Olmstead Map (fig. 3.47) showing some ancient shorelines related to the ancestral stages of Lake Erie, Ohio. (See Exercise 18C for exact location of the profile.)

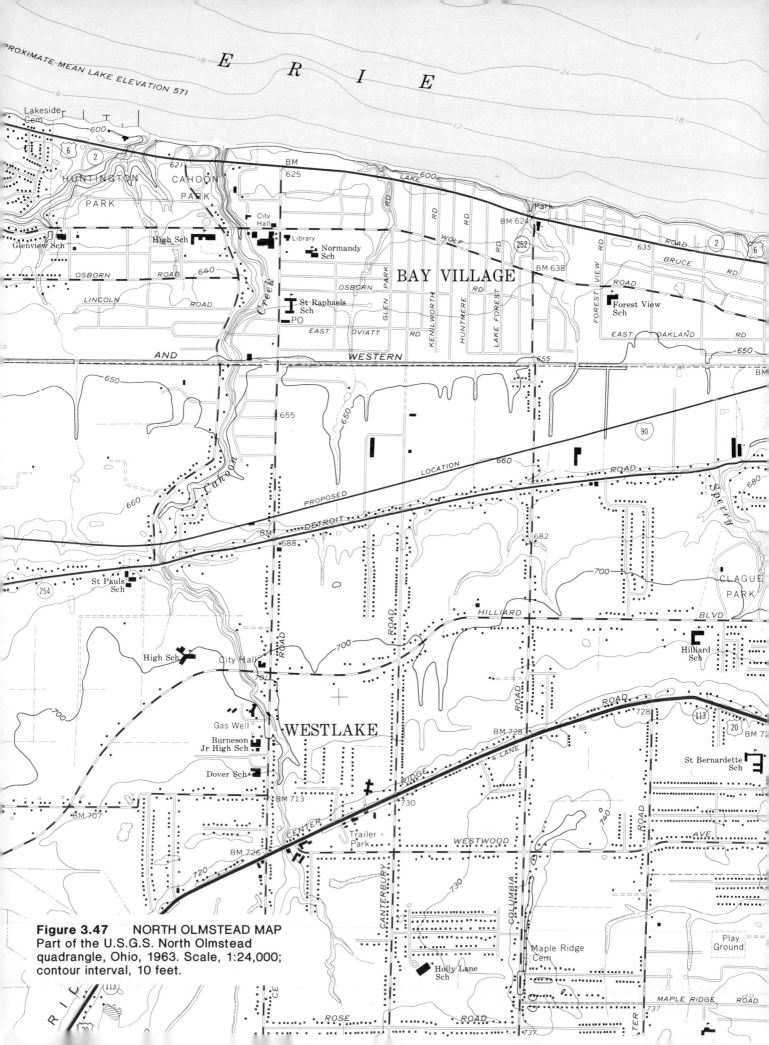

Figure 3.47 NORTH OLMSTEAD MAP
Part of the U.S.G.S. North Olmstead
quadrangle, Ohio, 1963. Scale, 1:24,000;
contour interval, 10 feet.

Exercise 18D. Shore Erosion and Levels of Lake Michigan

1. The record of levels for Lake Michigan (fig. 3.48) shows highs and lows over a period of 39 years. Does this record suggest that the fluctuations of the lake levels follow a regular periodicity that would permit the forecasting of future lake levels? Explain your answer.

2. What physical characteristics of the sediment exposed in the wave-cut cliff of figure 3.49 made this cliff particularly susceptible to the wave erosion during the high lake levels of 1986?

3. The chart of figure 3.48 shows a pronounced drop in the level of Lake Michigan between 1986 and 1989. Describe the impact of this drop on the sand bluff in figure 3.49.

4. The face of the bluff in figure 3.49 is littered with debris from the abandoned house perched on top. What additional signs are visible on the face of the bluff to indicate that it was undergoing severe erosion at the time the photograph was taken?

Reference

National Geographic, vol. 172, no. 1, July 1987. (A popular account of the hydrology of the Great Lakes.)

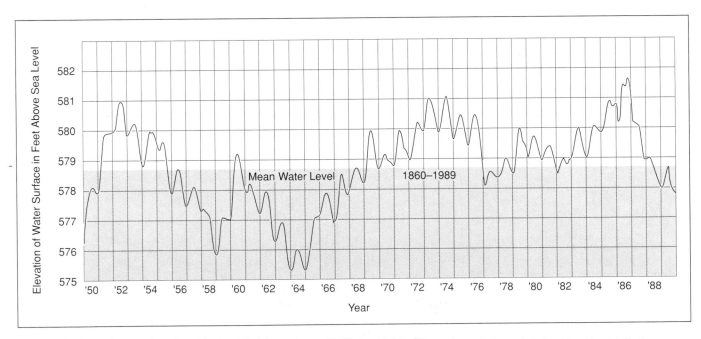

Figure 3.48   Water levels of Lake Michigan from 1950 to 1989. (Based on data published by the U.S. Army Corps of Engineers, Detroit, Michigan 48231.)

Figure 3.49 An ancient sand dune exposed to wave erosion on the shore of Lake Michigan near Muskegon, Michigan, November 1986. (Photo by Marge Beaver, Muskegon, Michigan.)

Landforms Produced by Volcanic Activity

Background

Over 500 active volcanoes occur on earth today, and thousands of extinct volcanoes are spread over the land surface and beneath the oceans. An *active volcano* is one that is in an eruptive phase, has erupted in the recent past, or is likely to erupt in the future. A *dormant volcano* is a term applied to an active volcano during periods of quiescence. An *extinct volcano* is one in which all volcanic activity has ceased permanently.

In this section we will deal with two of the major types of volcanoes, the shield volcano as exemplified by those on the island of Hawaii, and the composite or stratovolcano as represented by Mount St. Helens in the state of Washington. A *shield volcano* is a gently sloping dome built of thousands of highly fluid (low viscosity) lava flows of basaltic composition. A *composite volcano* or *stratovolcano* is a conical mountain with steep sides composed of interbedded layers of viscous lava and pyroclastic material. *Pyroclastic* refers to all kinds of clastic particles ejected from a volcano, the most ubiquitous of which is commonly called *volcanic ash*. The lavas in a composite volcano are rhyolite, dacite, or andesite. (Dacite is an aphanitic igneous rock intermediate in composition between rhyolite and andesite.)

Distribution of Volcanoes

Active volcanoes are widespread around the world, but their specific locations are controlled by conditions existing in the earth's crust. Composite volcanoes are concentrated along the margins of tectonic plates that abut against each other, and shield volcanoes occur over *hot spots,* deep-seated zones of intense heat whose geographic coordinates remain fixed for several millions of years. We will become more familiar with hot spots and tectonic plates in Part 5 of this manual. For purposes of the discussion here, it is sufficient to know that the zone formed where two plate margins converge toward each other is characterized by intense volcanic and earthquake activity. The most active convergent plate margin is the "Pacific Rim of Fire," part of which is shown in figure 3.50. Mount St. Helens is one of the many composite volcanoes lying on the North American segment of the Pacific Rim. The hot spot over which the island of Hawaii lies is not associated with a plate margin but occurs beneath the middle of the Pacific plate (fig. 3.50).

The shape and size of a shield volcano and a composite volcano are shown for comparison in the topographic profile of figure 3.51.

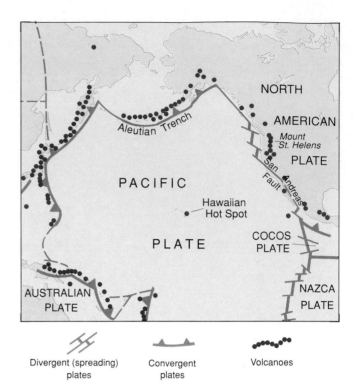

Figure 3.50 Map showing locations of the Hawaiian hot spot and Mount St. Helens with respect to part of the Pacific Plate and its margins. (From R. I. Tilling, et al., 1987. *Eruptions of Hawaiian Volcanoes: Past, Present, and Future.* U.S. Geological Survey.)

The Shield Volcanoes of Hawaii

The Island of Hawaii, also called the Big Island, is part of the Hawaiian Ridge, a partially submerged chain of volcanic mountains that rise from the deep ocean floor and extend from the Big Island to the northwest for a distance of more than 2,000 miles. All of the islands along the Hawaiian Ridge are shield volcanoes that were formed over the Hawaiian hot spot as the ocean floor moved progressively to the northwest over the last 40 million years.

The Big Island of Hawaii is the largest of the Hawaiian Islands (fig. 3.52) and consists of five shield volcanoes, of which only two, Mauna Loa and Kilauea, are active (fig. 3.53). Together, these five volcanoes rise 30,000 feet above the floor of the Pacific Ocean to an elevation more than 13,000 feet above sea level. The oldest of the five is Kohala, which became inactive about 60,000 years ago; Mauna Kea, the next oldest, ceased its eruptive activity about 3,000 years ago. Hualalai has erupted only once in historic times, during 1800–1801. Loihi, an active submarine volcano off the southeast coast of the Big Island, rises about 10,000 feet above the ocean floor to its summit 3,000 feet below the ocean surface. Loihi is the youngest volcano on the Hawaiian Ridge, and may become the next island in the Hawaiian Islands.

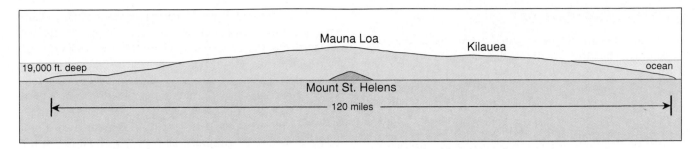

Figure 3.51 Topographic profile showing the comparison in size and shape of Mauna Loa and Mount St. Helens. (Modified from R. I. Tilling, et al., 1987. *Eruptions of Hawaiian Volcanoes: Past, Present, and Future.* U.S. Geological Survey.)

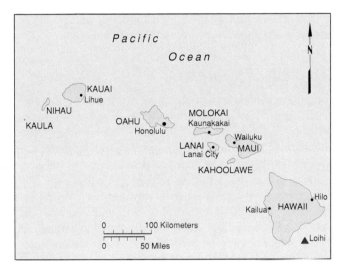

Figure 3.52 Map of the Hawaiian Islands in the Pacific Ocean. (From R. W. Decker, et al., eds., 1987. *Volcanism in Hawaii.* U.S.G.S. Professional Paper 1350.)

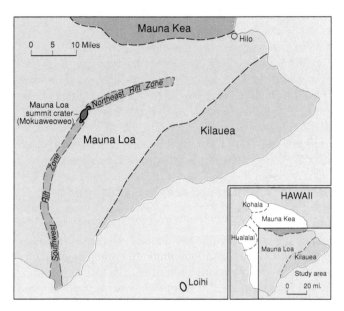

Figure 3.53 Map showing the location of the rift zones and the summit crater of Mauna Loa on the island of Hawaii. Inset shows the names and boundaries of the five shield volcanoes that form the architecture of the Big Island. (Modified from R. I. Tilling, et al., 1987. *Eruptions of Hawaiian Volcanoes: Past, Present, and Future.* U.S. Geological Survey.)

Hawaiian-Type Eruptions

Almost all Hawaiian-type volcanic eruptions on the Big Island during historic times (roughly the period from 1843 to the present) have occurred on Mauna Loa and Kilauea. These two volcanoes have been intensely studied by volcanologists in recent years, and it is from their observations that we are able to know a great deal about Hawaiian-type eruptive activity.

Hawaiian-type eruptions are weakly explosive or non-explosive. They extrude highly fluid basaltic lava that flows easily down the gentle slopes of the volcano's flanks. Indeed, it is the low viscosity of the lavas that accounts for the gentle slopes to begin with. Lava is erupted not only from the crest of the volcano but also from rift zones along its flanks. The rift zones on Mauna Loa are called the Southwest Rift Zone and the Northeast Rift Zone (fig. 3.53). The lava extruded from these rifts comes from a magma reservoir a few miles beneath the summit.

In the early stages of a Hawaiian-type eruption, lava spouts from fissures in the rift zones as *lava fountains.* These spectacular "curtains of fire" rise to heights of 100 feet or more; the largest lava fountain ever recorded rose to 1,900 feet on Kilauea in 1959. Lava derived from lava fountains or oozing from fissures flows downhill as incandescent rivers. These lava flows may reach speeds of 35 miles per hour before they congeal into solid rock. Two types of lava result, pahoehoe (pronounced "pa-hoy-hoy") and aa (pronounced "ah-ah"). *Pahoehoe* is commonly called "ropy lava" because it looks like a profusion of snarled ropes. *Aa* is a jagged, blocky mixture of irregular shapes and sizes. Eruptions produce both pahoehoe and aa lavas in widely varying proportions. Pahoehoe may turn into aa lava downstream from the point of extrusion, but aa flows do not change into pahoehoe.

Some lava flows reach the sea, where they plunge into the water and cool rapidly to form pillow lavas. *Pillow lavas* resemble a pile of pillows formed when chunks of molten lava, ranging in diameter from several inches to more than a foot, develop an outer solid skin and become draped over each other like a pile of water-filled balloons as they settle on the sea floor. Pillow lavas also form where lava is extruded directly into the sea through submarine vents. Most of the mass of shield volcanoes is believed to consist of pillow lavas.

During the infrequent explosive phases of Hawaiian-type volcanoes, lava and its contained gases are ejected violently into the atmosphere where the pieces solidify and fall back to the ground as *tephra,* a collective term assigned to all sizes of airborne volcanic ejecta. The constituent particles in tephra are called *pyroclasts,* a term which refers to any clastic particle derived from a volcanic eruption, be they preexisting rocks fragmented by a volcanic explosion or particles of lava that solidified while aloft. The smallest pyroclasts are collectively known as *volcanic ash.* Larger pyroclastic particles are referred to as cinders, scoria, pumice, and volcanic bombs. Cinders, scoria, and pumice are formed from rapid cooling of frothy lava from which gases were escaping during a fountain eruption. Pumice may contain so many gas-bubble cavities that it is light enough to float on water.

Pyroclastic materials account for only 1 percent of the mass of a Hawaiian-type volcano above sea level.

Composite Volcanoes

Composite volcanoes are steep-sided conical mountains consisting of flows formed from viscous ("sticky") lava interbedded with pyroclastics. Composite volcanoes erupt with great violence because the gases contained in the viscous lava build to a very high pressure before the stiff lava finally breaks through to the surface. The initial explosion blasts pulverized rock fragments several miles into the atmosphere. The smallest particles form volcanic ash that is carried by prevailing winds and deposited eventually in measurable thicknesses over hundreds of square miles downwind from the eruptive site. Volcanic ash produced by large explosive eruptions may be carried around the world.

A composite volcano may lie dormant for a hundred years or more before it is reactivated by an explosive eruption. Violent eruptions of composite volcanoes cause widespread destruction of flora and fauna, extensive damage to or total loss of property, and have been responsible for the loss of more than 200,000 lives since the dawn of civilization.

The violent eruption of Mt. St. Helens in the Cascade Range of the northwestern United States (fig. 3.54) in 1980 gave volcanologists an unprecedented opportunity to study the eruptive stages of a typical composite volcano, and much of what follows is based on their published accounts.

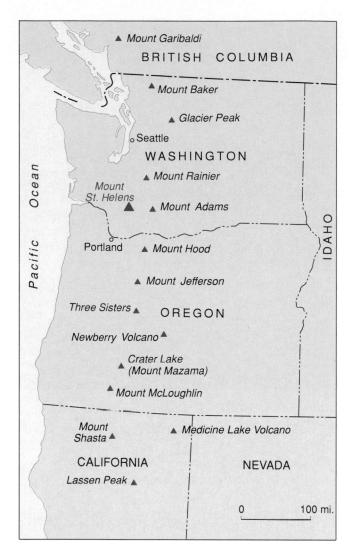

Figure 3.54 Map showing the location of Mount St. Helens and other composite volcanoes in the Cascade Range from northern California to southern British Columbia. The Cascade Range lies on a segment of the Pacific Rim where the Pacific tectonic plate is abutting against the North America plate. (From R. I. Tilling, 1987. *Eruptions of Mount St. Helens: Past, Present, and Future.* U.S. Geological Survey.)

Exercise 19A. Mauna Loa, a Hawaiian Shield Volcano

Figure 3.55 shows the surface distribution of lava flows produced from eruptions of Mauna Loa. The colored areas of the map represent lava flows aggregated into five age groups ranging from the youngest, the Historical flows formed between 1843 and the present, to the oldest, Group I, those formed more than 4,000 years ago. The ages of the flows that are contained in each group are based on radiocarbon-dated charcoal recovered from beneath the flows.

1. When was the youngest lava flow shown on the map deposited?

2. Why is the area of exposure of Groups I and II more widespread on the lower flanks of the volcano than near the summit or along the Southwest or Northeast Rift Zones?

3. Did any of the lavas in the five groups *not* reach the coastline?

4. The aa lavas are shown on the map by a stippled pattern. Aa lavas occur at any elevation on the volcanoes' flanks, but they are most common on the lower flanks. What is the reason for this?

5. The colored bands on the map that represent Historical and Group IV lavas generally are narrower than the colored bands representing lavas in Groups I, II, and III. Account for this difference.

6. Some of the lavas that reached the coast were pahoehoe (non-stippled areas), and some that reached the coast were aa. Why would a diver be more likely to encounter pillow lavas off the coast of pahoehoe lavas than off the coast of aa lavas?

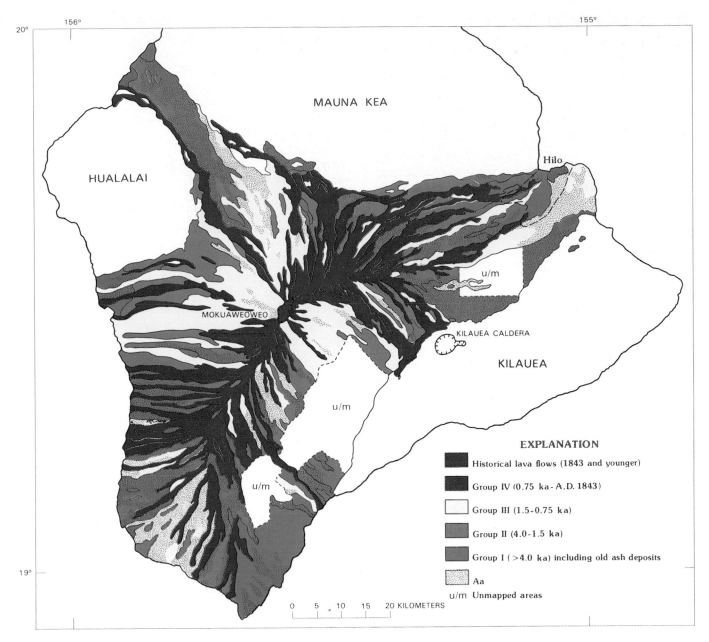

Figure 3.55   Map of Mauna Loa showing the surface distribution of lava flows in five different age categories. The notation "ka" stands for thousands of years before the year 1950. Thus, 0.75 ka = 750 years; 1.5 ka = 1,500 years; and 4.0 ka = 4,000 years. (J. P. Lockwood, and P. W. Lipman, 1987, *Holocene Eruption History of Mauna Loa Volcano*, Chapter 18 in R. W. Decker et al., editors, *Volcanism in Hawaii,* U.S.G.S. Prof. Paper 1350.)

The Eruption of Mount St. Helens

On May 18, 1980, Mt. St. Helens, one of the several spectacular volcanic peaks in the Cascade Range of the Pacific Northwest, underwent a violent volcanic eruption after two months of low-level activity characterized by earthquakes, steam venting, and small ash eruptions. This eruption of Mt. St. Helens, ending a 123-year dormant period, began with seismic activity on March 20, 1980. Steam vents opened up and ash began to be erupted by March 27. Small craters close to the summit developed, ash plumes were erupted, and ash avalanches took place in early April. Seismic activity continued, and by mid-April a significant crater had formed. The summit area of Mt. St. Helens began to swell, enough so that Goat Rocks on the northern flank had a measured movement of 20 feet vertically and 9 feet horizontally to the northwest.

As this activity continued, ash and steam continued to be produced, but in late April seismic activity was greatly reduced. In early May as swelling of the peak continued, the U.S. Geological Survey reported that the northern rim of the crater was rising at a rate of 2 to 4 feet per day.

Through the use of remote sensing techniques utilizing infrared film, hot spots were recorded in early May. Swelling continued, seismic activity increased, and at 8:32 A.M. on May 18, 1980, the violent eruption of Mt. St. Helens occurred, an event that claimed over 60 lives, devastated over 200 square miles of timberland and recreational areas, and spread measurable thickness of ash over several states, and eventually around the world.

The eruption itself was marked initially by an earthquake that triggered a major landslide down the north side of the mountain, followed quickly by a violent explosion. The two events combined to destroy the north rim of the summit crater (location 7, fig. 3.56). The rock material moved down the north side of the mountain as a mixture of rock, ash, steam, and glacial ice (location 6, fig. 3.56). Additional fluidization occurred when this avalanche mass hit the water of Spirit Lake and Toutle River. The velocity of the avalanche has been estimated at over 150 mph.

A portion of this enormous avalanche flowed down the valley of the Toutle River for 13 miles, depositing materials in a swath up to 1.2 miles wide and with a thickness up to 450 feet (fig. 3.56). Another part of the avalanche continued to the north, rose over a ridge that was 1,000 feet high, depositing over 100 feet of debris on top of the ridge before pouring over into the valley of South Coldwater Creek on the north side of the ridge (location 4, fig. 3.56). New lakes were formed as stream valleys were dammed by debris (location 3, fig. 3.56), and new islands were formed in Spirit Lake (location 5, fig. 3.56). The heavy black line on figure 3.56 marks the southern edge of the "Eruption Impact Area" as defined by the U.S. Geological Survey.

Large areas were covered by mudflows (unstippled gray areas, fig. 3.56) that resulted from the mixture of water from melting glaciers and large quantities of ash that poured out during the eruption. The major river draining the area to the north of Mt. St. Helens, the Toutle River, carried large quantities of sediment almost as mudflow to the west into the Cowlitz River and eventually into the Columbia River. Previously recorded flood stages on the Toutle River were exceeded by almost 30 feet. Silting occurred in the Columbia River at the mouth of the Cowlitz River, trapping ships upstream. Dredging of a channel was necessary before these ships could move down river to the ocean.

Ash falls occurred to the east of Mt. St. Helens affecting cities such as Yakima, in the heart of the apple growing district of Washington, and the major wheat growing areas farther east. At Ritzville, 205 miles east of Mt. St. Helens, a fine ash deposit of 70 mm was recorded, and by May 21 ash had spread across the continent to the east coast. A later ash eruption on May 25 spread ash to the northwest, mantling an area lying roughly between the Columbia River and Olympia, Washington. Small ash eruptions have occurred since, but none as large as the May 25 activity.

The eruption of Mt. St. Helens was the first such volcanic event in the contiguous 48 states since the eruption of Mt. Lassen in northern California that began in 1914 and continued until 1921. Volcanic activity in the Cascade Range was recorded during the 1800s, and several peaks such as Mt. Rainier, Mt. Baker, and Mt. Hood still have active fumaroles.

Volcanic activity has continued on Mt. St. Helens at a much reduced rate in the years following the 1980 eruption. This activity has included gas and ash emissions, earthquakes, rockfalls, and extrusion of a lava spine in the crater, and there is now forming a "composite dome" in the center of the crater. This activity is continually monitored by the U.S. Geological Survey and is providing considerable information that may help in the development of earthquake and volcanic eruption prediction models.

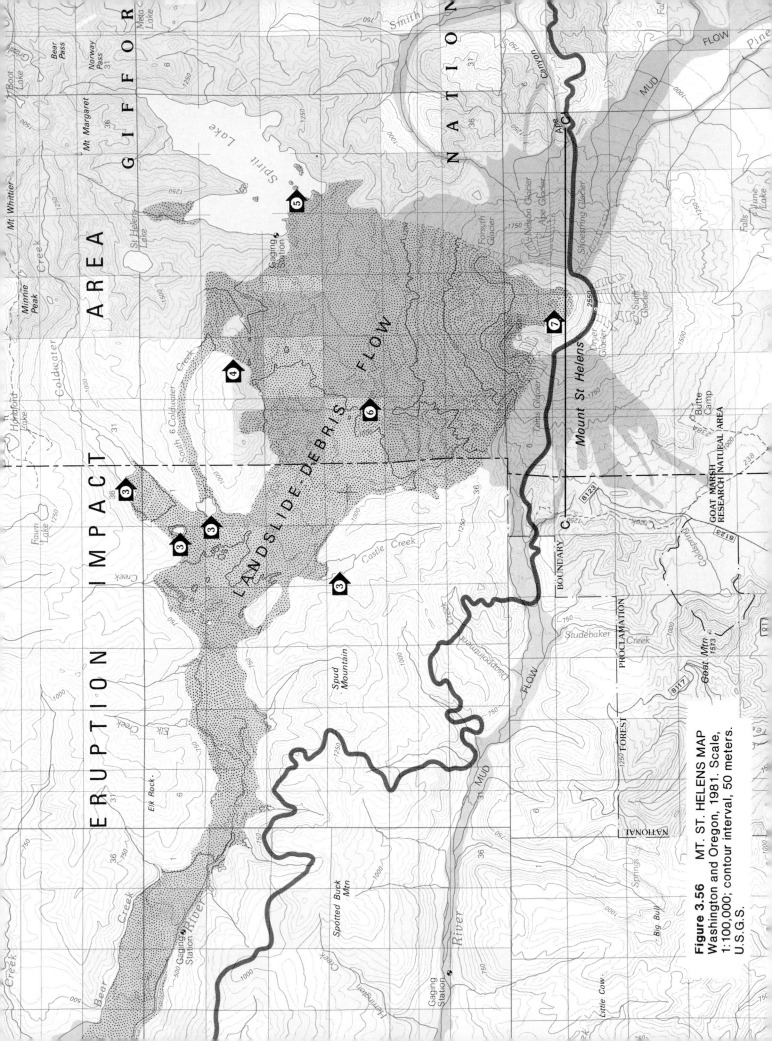

Figure 3.56 MT. ST. HELENS MAP
Washington and Oregon, 1981. Scale,
1:100,000; contour interval, 50 meters.
U.S.G.S.

Exercise 19B. The Impact of the Eruption of Mount St. Helens on Surrounding Topography

The eruption of Mt. St. Helens resulted in several changes in the topography of the Eruption Impact Area, the southern boundary of which is shown as a heavy black line in figure 3.56. Figure 3.56 is a post-eruption map published in 1981, and figure 3.57 is a map of the same area made before the eruption and published in 1980. The pre-eruption map will be referred to as the "Old Map," and the post-eruption map will be called the "New Map." Notice that the contour interval of both maps is 50 meters.

1. The elevation of Spirit Lake was 974.7 m on the Old Map. Determine the elevation of the contour line nearest the shoreline of Spirit Lake on the New Map, and answer the following questions:
   a) Was the elevation of Spirit Lake increased or decreased because of the eruption?
   b) Was the area of Spirit Lake increased or decreased because of the eruption?
   c) Estimate the *maximum* amount of increase or decrease in the elevation of Spirit Lake in *feet*.

2. The North Fork of Toutle River was the outlet of Spirit Lake on the Old Map. Why is there no outlet of Spirit Lake shown on the New Map?

3. What impact did the eruption have on the *valley* of the North Fork of the Toutle River?

4. Draw in blue pencil on the Old Map the outlines of the lakes shown on the New Map that occur along the courses of Coldwater Creek, South Coldwater Creek, and Castle Creek. How did these lakes come into existence?

5. The flanks of Mt. St. Helens are draped with alpine glaciers on the Old Map. How have the area and length of these glaciers been changed by the eruption?

6. The shape of Mt. St. Helens is a nearly symmetric cone on the Old Map with a summit elevation of 2,950 m. The elevation of the summit of Mt. St. Helens on the New Map is 2,550 m, and it lies at a different position than it did before the eruption. Trace the 2,250-meter contour line around the top of the mountain on both maps to show how the mountain top changed because of the eruption. Explain why the summit elevation is lower on the New Map and why it has changed positions from the pre-eruptive summit.

7. The landslide-debris flow shown on the New Map altered the terrain over which it moved. Would you describe the *new* landscape as a constructional or destructional surface? Explain why you called it one or the other.

References

Tilling, R. I., Heliker, C., and Wright, T. L., 1987, *Eruptions of Hawaiian Volcanoes: Past, Present, and Future*, U.S. Geological Survey.

Tilling, R. I., 1987, *Eruptions of Mount St. Helens: Past, Present, and Future*, U.S. Geological Survey.

The material presented in this section of this Laboratory Manual was drawn freely from these two outstanding publications. They are both excellent sources of information on the eruptions of the Hawaiian-type and composite volcanoes.

Figure 3.57 MT. ST. HELENS MAP
Washington and Oregon, 1980. Scale, 1:100,000; contour interval, 50 meters. U.S.G.S.

Structural Geology

Part 4

Background

Structural geology deals with the architectural patterns of rock masses as they occur in nature. In Part 1 you were introduced to some basic concepts and principles about the occurrence of rocks. In Part 4, we will build on the concepts and related terminology introduced in Part 1. It may be useful for you to review the text and diagrams on pages 40–47.

Structural geology involves all three rock types—igneous, sedimentary, and metamorphic—but in this part of the manual we concentrate on structures in which sedimentary rock layers are dominant. A sedimentary rock unit that is characterized by a distinct lithologic composition is called a *formation*. A formation is the basic stratigraphic unit depicted on a geologic map. The boundary between two contiguous formations is called a *contact*.

For the sake of simplicity, the formations illustrated herein will be homogeneous in their lithology and will be referred to by some name such as "limestone formation," "sandstone formation," or some other name. In reality, however, a formation may consist of several thinner layers or beds. The contacts between these beds are called *bedding planes* and are more or less parallel to the contacts of the formation itself.

It is common geologic practice to assign a name to a formation. In some cases, the name of the formation will include a lithologic descriptor such as the Madison Limestone or the Pierre Shale. In other cases, the name will consist only of a proper name such as the Nelson Formation or the Sunflower Formation. Formational names such as these will be found only on published geologic maps. On simple diagrammatic maps that are used herein, it will suffice to use only a lithologic definition such as the shale formation or sandstone formation.

Sedimentary strata occur in a variety of three-dimensional geometric forms. The pattern of contacts between sedimentary formations portrayed on a map shows only the distribution of the formations in two dimensions. The visualization of a three-dimensional geometric form from a two-dimensional geologic map is one of the main objectives of this part of the manual. By learning certain basic principles and following standard procedures described in the pages that follow, the ability to formulate a mental three-dimensional picture of the geometric configuration of strata beneath the earth's surface can be mastered.

Geologic structures are produced when strata are deformed by forces that contort them from their original position of horizontality into geometric forms called *folds*. Folded rock layers retain their continuity as layers, that is, the contact between two formations on a geologic map can be followed as an unbroken line. When exposed by erosion, these folds are revealed in a two-dimensional *outcrop pattern* on a geologic map that is diagnostic of the three-dimensional forms of the structures.

In many cases, deformation causes the strata to break or *fault* along *fault planes* so that the outcrop pattern shows a discontinuity of the formations on either side of the fault. Movement along a fault plane produces an *earthquake*.

Folds, faults, and earthquakes therefore are the subject of Part 4, and will be treated in that order in the pages that follow.

Structural Features of Sedimentary Rocks

Deformation of Sedimentary Strata

During the course of geologic history, sedimentary strata have been subjected to vertical and horizontal forces that may alter the original horizontal position of the rock layers. Some strata may be uplifted in a vertical direction only, so that their original horizontality remains more or less intact. In other cases, the forces of deformation produce architectural patterns ranging from simple to extremely complex structures.

In order to decipher these structures, geologists measure certain features of a given formation where it crops out at the surface of the earth. These measurements define the position of the formation with respect to a horizontal plane of reference. The precise orientation of a contact, bedding plane, or any planar feature associated with a rock mass is called the *attitude*. When attitudes from many outcrops are plotted on a base map such as a topographic map or aerial photograph, and combined with the contacts between formations, the overall geometric pattern or structural configuration of the strata can be determined.

Components of Attitude

The attitude of a formation consists of three parts that collectively define its position at a given location with respect to a horizontal plane and a compass direction.

1. *Strike:* A horizontal line in the plane of the bedding expressed as a compass direction.
2. *Direction of Dip:* The compass direction in which the layer is inclined downward from the horizontal. The direction of dip is always at right angles to the direction of the strike.
3. *Angle of Dip:* The angle between a horizontal plane and a bedding plane. The dip angle is measured in degrees.

The three components of attitude are shown in figure 4.1. As an example of a verbal description of the attitude of a formation at a particular site, the following notation would be used: At the north end of the bridge across the Snake

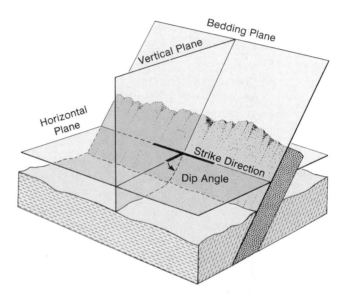

Figure 4.1   Three-dimensional view of an outcropping of sandstone in which the attitude of a bedding plane is measured with respect to horizontal and vertical planes. The shaded slanting plane represents the bedding plane of the layered sandstone. The intersection of the bedding plane and a horizontal plane results in a line called the *strike* of the formation. On a map this line is expressed as a compass direction. The angle formed by the horizontal plane and the bedding plane is the *dip* of the formation. The dip angle is always measured in a vertical plane that is perpendicular to the direction of strike. (From J. H. Zumberge and C. A. Nelson. *Elements of Physical Geology.* Copyright © 1976 New York: John Wiley & Sons, Inc., New York.)

River, the Sundance Formation strikes north 45 degrees east and dips to the southeast at an angle of 30 degrees. On a geologic map, however, the attitude of a formation would be shown by a *strike and dip symbol.* Various forms of this symbol are given in figure 4.2. In illustrations used in parts of this manual, the strike and dip symbols may appear without the notation of the angle of dip.

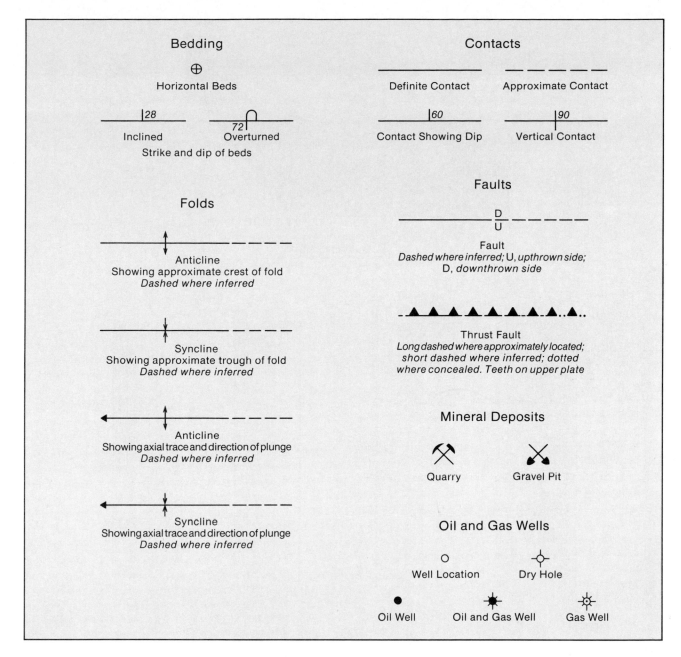

Figure 4.2 Standard symbols used on geologic maps.

Methods of Geologic Illustration

Geologic information gathered by the study of outcrops is displayed in a number of ways in order to depict the overall structural features and relative age relations of the strata involved. The three main types of geologic illustrations or diagrams are as follows.

1. *Geologic Map:* This is a map that shows the distribution of geologic formations. Contacts between formations appear as lines, and the formations themselves are differentiated by various colors. The map may also show topography by standard contour lines.

2. *Geologic Cross Section:* A diagram in which the geologic formations and other pertinent geologic information are shown in a vertical section. It may also show a topographic profile, or it may be schematic and show a flat ground surface.

3. *Block Diagram:* A perspective drawing in which the information on a geologic map and geologic cross section are combined. This mode of geologic illustration is used to show the three-dimensional aspects of a geologic structure.

Figure 4.3 shows how these three methods of geologic illustrations are related.

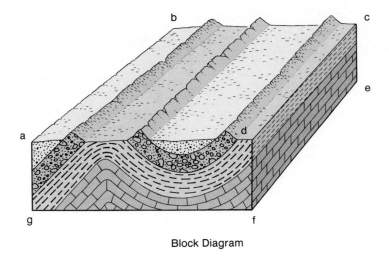

Block Diagram

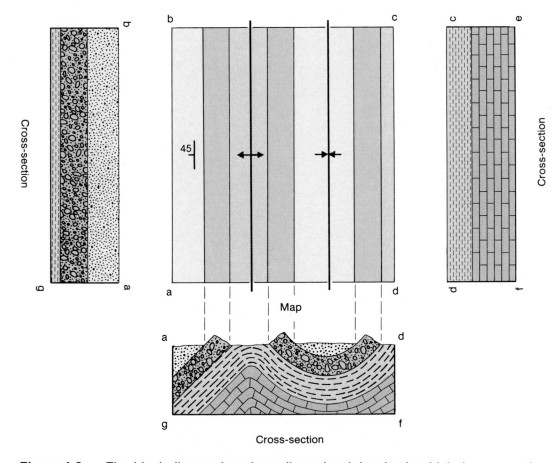

Figure 4.3 The *block diagram* is a three-dimensional drawing in which the geometric configuration of the geologic structure is depicted. A *geologic map* shows the areal extent of formations at the earth's surface and contains certain symbols that further define the geometry of the rock masses as they extend beneath the surface. A *geologic cross section* is a view of the geologic formations in a vertical plane.

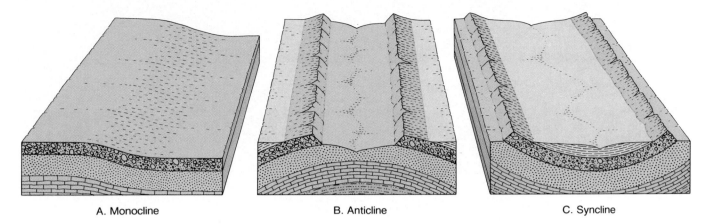

A. Monocline   B. Anticline   C. Syncline

Figure 4.4 Block diagrams of three common folds. The dissected ridges formed by resistant layers in diagrams *B* and *C* are called hogback ridges.

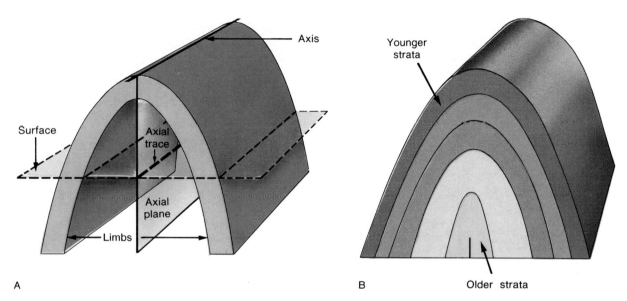

Figure 4.5 (*A*) Nomenclature of a fold. (*B*) Age relationships of strata in an anticline. (From Carla W. Montgomery, *Physical Geology,* 2d ed. Copyright © 1990 Wm. C. Brown Publishers, Dubuque, Iowa. All Rights Reserved. Reprinted by permission.)

Sedimentary Rock Structures

Sedimentary strata that have been subjected to forces of deformation result in three basic geologic structures as shown in the block diagrams of figure 4.4.

A. *Monocline:* A structure in which the strata have a uniform direction of strike but a variable angle of dip.
B. *Anticline:* A structure in the form of an arch.
C. *Syncline:* A structure in the form of a trough.

Notice that in figure 4.4 the arch of the anticline is not reflected in a corresponding topographic arch, and that the synclinal trough is a geologic trough, not a topographic one. The surface topography of the parallel ridges in figure 4.4*B* and 4.4*C* is controlled by a formation that is more resistant to erosion than the other formations in the structure.

Geometry of Folds

The geometry of a fold is more precisely defined by the attitude of the *axial plane* of the fold, an imaginary plane that separates the *limbs* of the folds into two parts as shown in figure 4.5*A*. The *axial trace* of the fold appears as a line on a geologic map.

If the axial plane is essentially vertical, the fold is said to be *symmetric* (fig. 4.6*A*); if the axial plane is inclined so that the limbs dip in opposite directions but one limb is steeper than the other, the fold is *asymmetric* (fig. 4.6*B*); and if the axial plane is inclined to the extent that the opposite limbs dip in the same direction, the fold is *overturned* (fig. 4.6*C*). A *recumbent fold* is an overturned fold in which the axial plane is nearly horizontal. The symbols used on geologic maps to show the traces of axial planes are shown in figure 4.2.

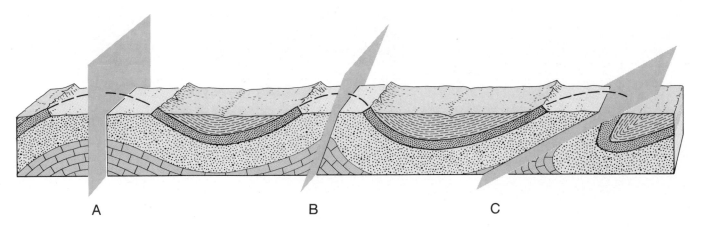

Figure 4.6 Block diagram in which three variations of a fold are shown: (*A*) symmetric anticline; (*B*) asymmetric anticline; (*C*) overturned anticline. Note the different attitudes of the three axial planes.

The folds shown in figures 4.4, 4.5, and 4.6 are *non-plunging folds* because the strikes of the limbs are parallel. Another way of describing a nonplunging fold is to say that the strikes of the folded formations are all parallel as shown in figure 4.3.

If, however, the strikes of the formations on either side of an axial plane converge, as in figure 4.7, the fold is said to be a *plunging fold.* A geologic map on which a series of plunging folds is displayed shows a *zig-zag* outcrop pattern. The *direction of plunge* is shown by an arrow placed on the trace of the axial plane, as in figure 4.7. *The direction of plunge of a plunging anticline is toward the apex of the converging formations* as seen on a geologic map, and *the direction of plunge of a plunging syncline is toward the open end of the V-shaped pattern of diverging formations.* Figure 4.7 shows both cases.

An anticline that plunges in opposite directions is a *doubly-plunging anticline,* and a syncline that plunges in opposite directions is a *doubly-plunging syncline.* Variations of doubly-plunging folds are the *structural dome* and *structural basin,* as shown in figure 4.8. The outcrop patterns of these two structures are more or less concentric circles.

Geologic Maps and Cross Sections

Geologic Maps

A geologic map shows the distribution of rock types in an area. The map is constructed by plotting strikes and dips of formations and the contacts between formations on a base map or aerial photograph. This information is based on field observations on outcrops in the map area. Because a single isolated outcrop rarely yields sufficient information from which the overall structural pattern for a given area can be understood, the geologist must visit enough outcrops in the map area to permit the filling of the gaps from one outcrop to another.

We are not concerned here with the making of actual geologic maps, but rather with their interpretation and the construction of geologic cross-sections from them. The interpretation of a geologic map requires an understanding of the information shown on them and the ability to translate that information onto a geologic cross section. To accomplish this, one must learn to visualize three-dimensional relationships from a two-dimensional pattern of geologic formations as they appear on a geologic map. This is perhaps the most difficult aspect of structural geology for a beginning student to master, but by following step-by-step instructions, these relationships eventually will become clear.

In general, keep in mind that a geologic map shows the distribution of formations as they appear at the surface of the earth. How this surface information can be used to visualize the unseen components of the rocks below the surface constitutes the subject matter to follow.

Having already been exposed to the notations and symbols used on a geologic map, you may find it useful to review the relationship between a geologic map and a geologic cross-section as shown in figure 4.3. If you thoroughly understand how the geologic cross section of figure 4.3 relates to the geologic map, and how both relate to the block diagram, you will be in a good position to proceed with the next steps in map interpretation.

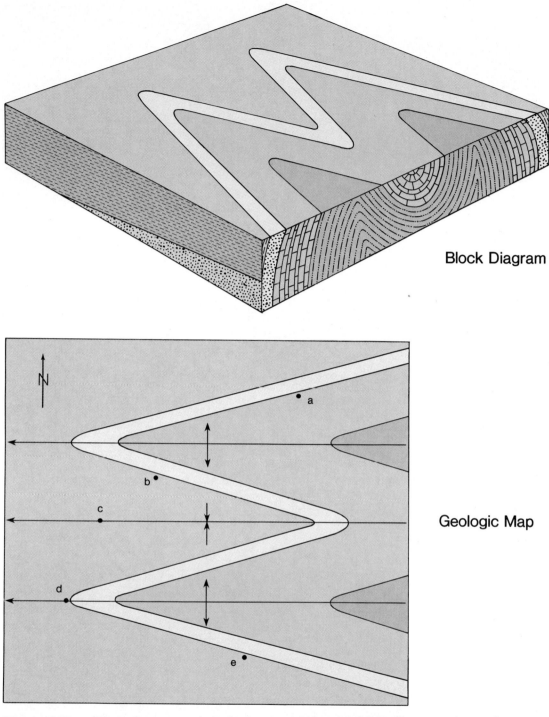

Block Diagram

Geologic Map

Figure 4.7 Block diagram and geologic map of plunging folds. The map shows the characteristic outcrop pattern of plunging folds. Here, two plunging anticlines and one syncline plunge to the west. If this area were in a humid region, the arkose formation would be more resistant to erosion than the shale, limestone, or siltstone formations and would therefore form a hogback ridge.

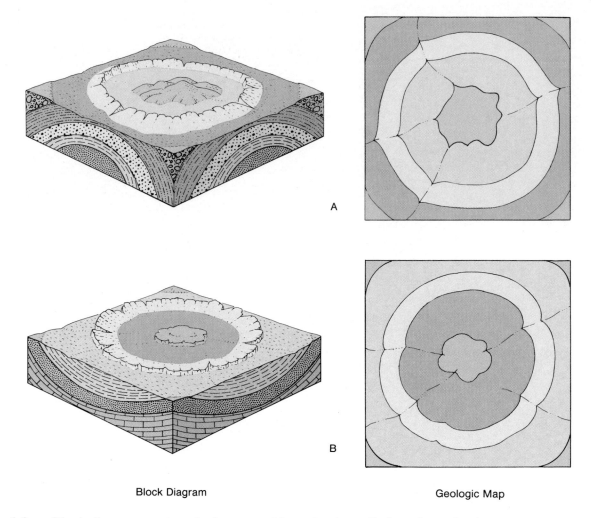

Block Diagram   Geologic Map

Figure 4.8   Block diagrams and geologic maps of two structures that produce circular
outcrop patterns: (*A*) structural dome; (*B*) structural basin.

Geologic Cross Sections

The purpose of a geologic cross section is to display geologic features in a vertical section, perpendicular to the
ground surface. A crude but nonetheless accurate analogy
is to be found in the ordinary layer cake. When a layer
cake is viewed from above, all that can be seen is the
frosting; the "structure" of the cake is obscured. However, if the cake is cut vertically and the two halves are
separated, the component layers of the cake constitute a
cross section of the cake so that its structure will be revealed.

Geologic formations do occur in "layer-cake" structures, but they commonly occur in much more complex
structures, and it is through the construction of a geologic
cross section that these complexities are unraveled. Following are some general rules and guidelines for use in
constructing a geologic cross section from a geologic map.

1. A geologic cross section is constructed on a vertical
   plane. The cross section is shown on the corresponding geologic map by a line that is equivalent to
   the line along which the cake was cut in the layer-
   cake analogy. Information on or near the line of the
   cross section on the map is transferred to the cross
   section as the first step in its construction. Such notations as directions and angles of dip, formational
   contacts, traces of axial planes, and the like provide
   the basic elements used to make a geologic cross
   section.
2. Sedimentary formations to be drawn on cross
   sections in the exercises in this manual are assumed
   to have a constant thickness. That is to say, they do
   not thicken or thin with depth or along the strike.
3. Dip angles from strike and dip symbols on the map
   can be used as a basis for estimating the inclination
   of strata on a cross section. If dip angles are not shown,
   keep the dip angles as small as possible but consistent
   with the thickness of the strata and structural relationships.
4. The relative ages of sedimentary strata in some of the
   maps and cross sections used herein are designated by
   arabic numerals. For example, if four formations are

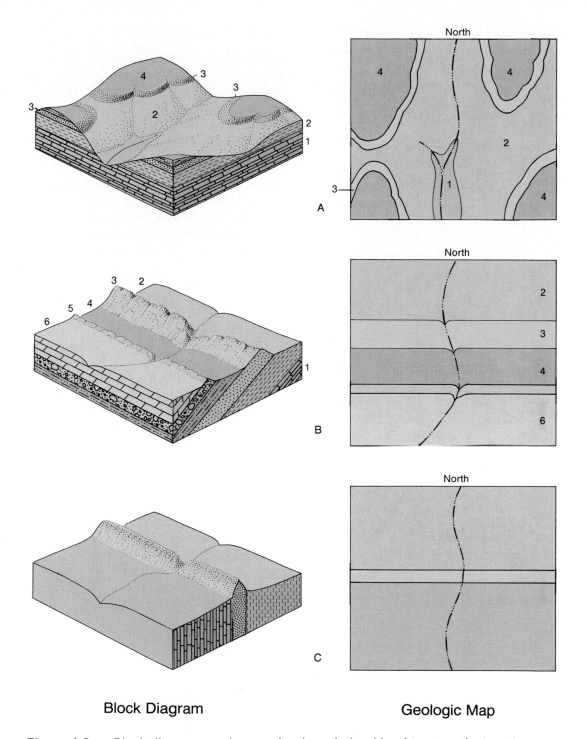

North

North

North

Block Diagram

Geologic Map

Figure 4.9 Block diagrams and maps showing relationship of topography to outcrop patterns. (*A*) Horizontal strata dissected by a drainage system. Numbers refer to relative ages of the formations. The formation labeled 1 is the oldest. (*B*) Tilted rock layers showing the V formed where a contact crosses a stream. The oldest beds (i.e., 1, 2, and 3) dip toward the youngest beds (i.e., 5 and 6). The apex of the V's points in the direction of dip. (*C*) Three vertical sedimentary beds, one of which is more resistant to erosion than the other two. In this case the law of V's cannot be used because no V's are formed. The relative age relationship of the three formations cannot be determined from the information shown either on the block diagram or on the map.

shown on a map or block diagram, the oldest formation is assigned the number "1," and the youngest, a number "4" (fig. 4.9A).

5. If you are required to draw a geologic cross section from a geologic map on which no strike and dip symbols are present, the direction of dip can be determined in the following manner.

a) Where a formation contact crosses a stream on the map it forms a V, the apex of which points in the direction of dip as shown in the geologic map of figure 4.9B. (This rule is not to be confused with the "law of V's" as applied to contour lines when they cross a stream.)

b) The shape of a V formed by a contact that crosses a stream may be used to estimate the angle of dip of the contact. A broad open V indicates a steep dip, and a narrow V is indicative of a shallow dip angle. Where no V is formed, the formation contact is vertical, as shown in figure 4.9C. The foregoing method for determining the direction of dip takes precedence over the method described next.

c) In a sequence of formations, none of which has been overturned, the oldest beds dip toward the youngest, as shown in the geologic map of figure 4.9B.

Width of Outcrop

When folded strata are exposed to erosion at the earth's surface, they appear as bands on the geologic map. The width of a single band is called the *width of outcrop* although the full thickness of the formation may not be exposed in a single outcrop. The width of outcrop is controlled by three factors: the thickness of the formation, the angle of dip of the formation, and the slope of the land surface where the outcrop is exposed.

To illustrate these controlling factors in the simplest case, consider the three horizontal formations of equal thickness in figure 4.10. The geologic cross section in figure 4.10A shows how the thickness of each formation varies with the slope of the land surface. A gentle slope results in a width of outcrop that is greater than the thickness of the formation, as in the case of the shale formation; and a steeper slope produces a width of outcrop that is less than the thickness of the formation, as in the case of the sandstone and limestone formations.

Two other cases of the relationship of thickness to the width of outcrop are shown in figure 4.10B and C. In figure 4.10B, where the beds are dipping 30 degrees, the thickness of each formation is shown on the cross section and the corresponding width of outcrop is shown on the geologic map. In figure 4.10C, the formations are vertical, that is, they dip 90 degrees. In this case, the true thickness of a formation is the same as the width of outcrop. The general rule, however, is that *the width of outcrop on a geologic map is not necessarily the same as the true thickness of the formation as seen in a geologic cross section.* This rule must be kept in mind when drawing cross sections from a geologic map or a block diagram in the exercise that follows.

Figure 4.10 Cross section and geologic maps of three sedimentary formations showing the relationship of the true thickness of each formation (T_{ss}, T_{sh}, and T_{ls}) to their widths of outcrop on a geologic map. The thickness of each formation is the same on all three cross sections. The width of outcrop (W_{ss}, W_{sh}, and W_{ls}) on the geologic map depends on the thickness and attitude of the formations and the slope of the surface. In the three cases shown, (*A*) horizontal strata, (*B*) inclined strata, and (*C*) vertical strata, only in *C* does the thickness of each formation in the cross section equal the width of outcrop of the same formation on the geologic map.

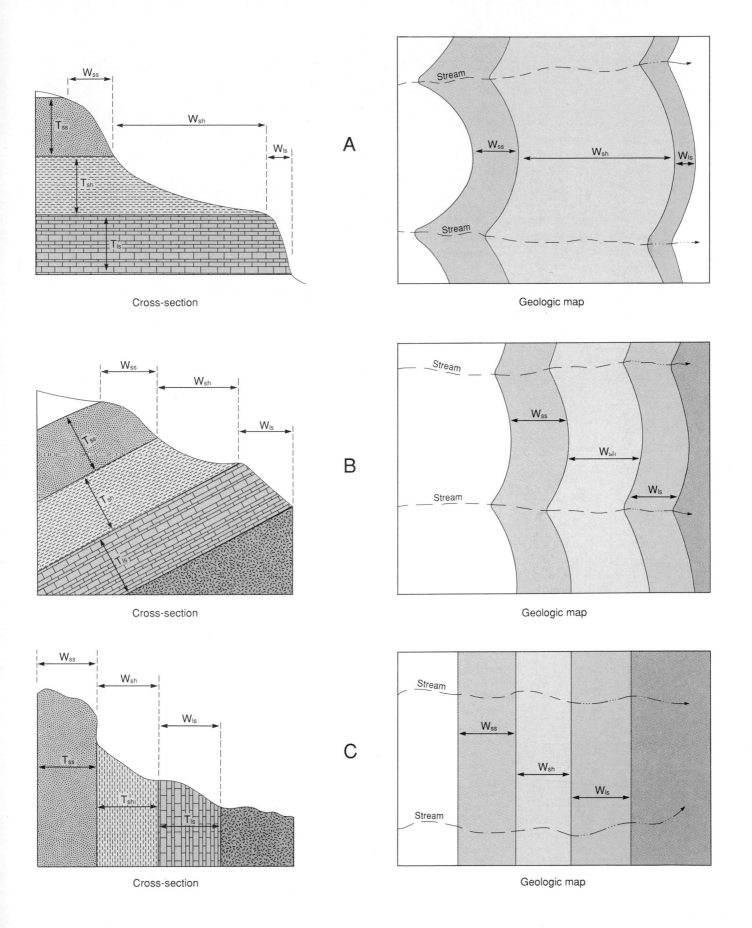

A

Cross-section

Geologic map

B

Cross-section

Geologic map

C

Cross-section

Geologic map

Exercise 20. Geologic Structures on Block Diagrams, Geologic Symbols, and Relative Ages of Formations

1. Complete the four block diagrams in figure 4.11. Below each block diagram, print the name of one or more of the geologic structures shown. Remember that the numbers on the map indicate the relative ages of the formations, with number 1 being the oldest. Assume that the topography in all four diagrams is essentially flat except in diagrams C and D where a stream cuts across the formational contacts.

2. On figure 4.7, place a strike and dip symbol at points *a, b, c, d,* and *e.* Use a black pencil.

3. Figure 4.8 shows two geologic maps. All formations shown there are sedimentary in origin.

   a) Label each formation with a number indicating its relative age in the sequence of strata. The oldest should be labeled number 1.

   b) Draw strike and dip symbols on each map.

4. On the geologic maps of figure 4.10*A* and *B,* draw the appropriate geologic symbol that shows the attitude of each of the three formations, and show by numbers the relative age of each.

5. Why is it impossible to tell the relative ages of the formations shown in figure 4.10*C* without reference to figure 4.10*A* or *B*?

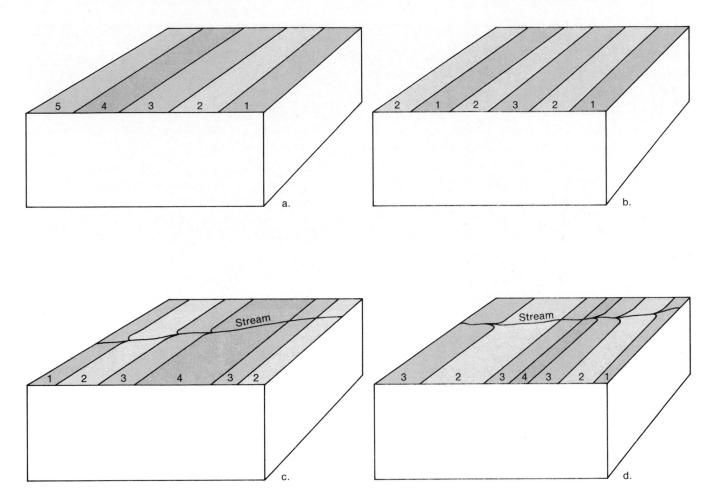

Figure 4.11   In these four block diagrams, dip directions in *A* and *B* are to be determined by the rule that older beds dip toward younger beds, and in *C* and *D*, dip directions are to be determined by the rule of V's.

Exercise 21. Geologic Mapping on Aerial Photographs

21A. CHASE COUNTY, Kansas

Figure 4.12 shows the outcrop pattern of sedimentary strata. The light and dark gray tones correspond to different lithologic characteristics of the various formations. At *A*, near the center of the photograph, the contacts of a light gray formation are shown by two black lines.

1. Extend these lines along the contacts as far as possible on the photo. Do the same for formation *B* shown in the upper left-hand corner of the photo.

2. What is the general attitude of these two formations? (Compare the outcrop pattern of fig. 4.12 with fig. 4.9*A*.)

3. If it is assumed that the thickness of these two formations is constant, why do their widths of outcrop change from place to place?

4. Applying the Law of Superposition in this case, which of the two formations is relatively older than the other?

Figure 4.12   Aerial photograph, Chase County, Kansas. Scale, 1:20,000. (U.S. Dept. of Agriculture photograph.)

21B. FREMONT COUNTY, Wyoming

This stereopair (fig. 4.13) shows sedimentary strata cropping out in the area. Study the stereopair with a stereoscope, and using figure 4.9B for guidance, answer the following questions.

1. Draw several strike and dip symbols on the right-hand photo of the stereopair, and write a verbal description of the attitude of the formations.

2. Are the oldest beds in the northern or southern part of the area? What rule is applied here that allows you to answer this question?

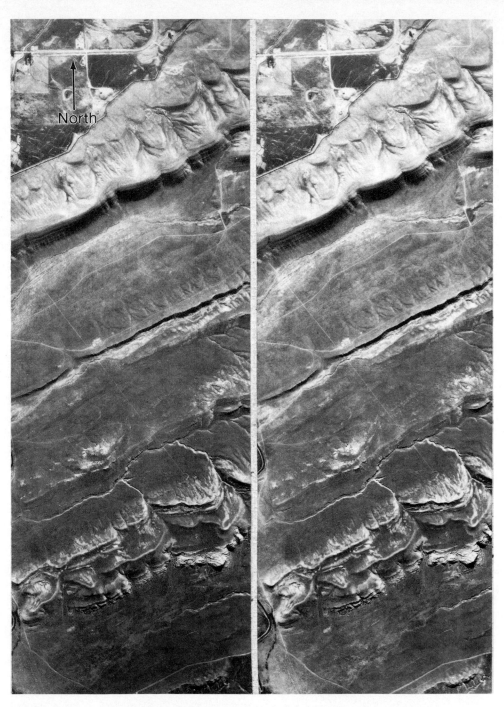

Figure 4.13   Stereopair of part of Fremont County, Wyoming. Scale, 1:21,500; July 13, 1960. (U.S.G.S.)

21C. AERIAL PHOTOGRAPH, Arkansas

The pattern of curved ridges in this photograph (fig. 4.14) is the result of differential erosion of sedimentary strata. The ridges are composed of rocks that are more resistant to erosion than the rocks that form the intervening valleys. The structure displayed in the photograph is the nose of a steeply plunging anticline.

1. Draw the trace of the axial plane of the fold on the photograph with a red pencil, and add other appropriate symbols on the axial trace and elsewhere on the photograph to indicate all relevant structural information. (Refer to fig. 4.2 as a reminder of the appropriate symbols to use for the fold axis.)

Figure 4.14   Aerial photograph showing the nose of a plunging anticline in Arkansas. Scale, 1:24,000; November 9, 1957. (U.S.G.S.)

21D. LITTLE DOME, Wyoming

The structure shown here (fig. 4.15) is an elongate dome or a doubly-plunging anticline. Use a stereoscope to study the stereopair while formulating the answers to the following questions.

1. Draw strike and dip symbols on the right-hand photograph of the stereopair. Use red pencil.

2. Draw the trace of the axial plane and other symbols that are appropriate for this structure. Draw on the right-hand photograph and use red pencil.

3. What is the evidence that the angle of dip changes as one follows the ridges northward along the eastern flank of the structure?

4. If a hole were drilled on the axis of the fold at the center of the structure, would the drill encounter any of the formations that crop out on the surface in the area covered by the photographs? Explain how you arrive at your answer.

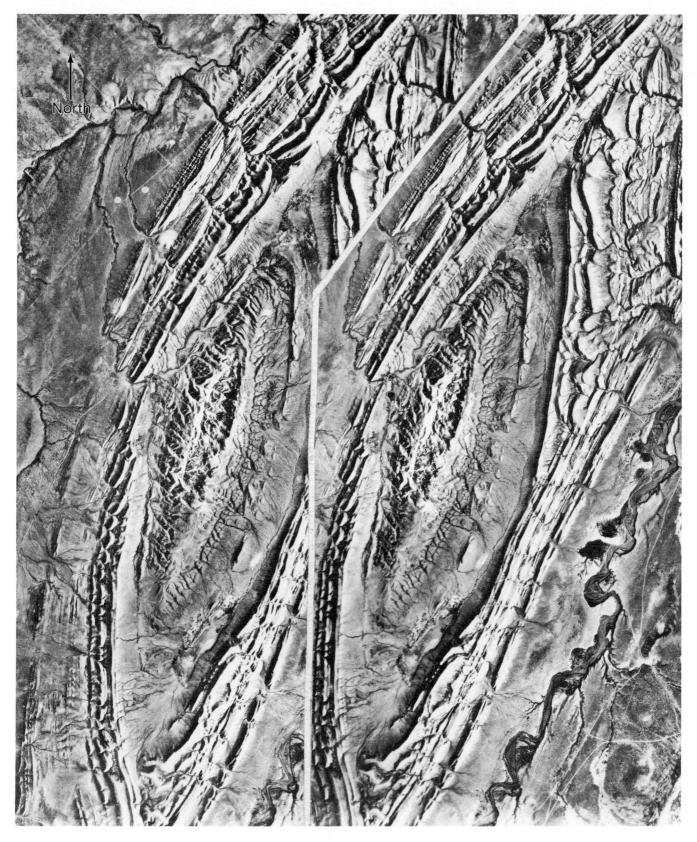

Figure 4.15   Stereopair of Little Dome, Wyoming. Scale, 1:23,600; October 20, 1948. (U.S.G.S.)

21E. HARRISBURG, Pennsylvania

Figure 4.16 looks like an aerial photograph, but in fact is a Side-Looking Airborne Radar (SLAR) image that enhances surface features. A SLAR image is particularly useful in deciphering the structural geology of an area.

The patterns of ridges and valleys shown here are the result of differential erosion of a series of plunging folds in the Appalachian Mountains of Pennsylvania. The river flowing through the area is the Susquehanna River, and the city of Harrisburg lies at the upper margin of the figure near the river.

Figure 4.16 is oriented so that the north direction is toward the lower margin of the figure. The reason for this is that the "shadows" produced by the imaging process must be toward the viewer in order for the ridges on the ground to appear as ridges on the image. (By turning the page upside down, you may see a "reverse topography," that is, the ridges appear as valleys.) Examine the image to get a feel for the structural pattern it reveals.

1. Locate the zig-zag ridge that is cut by the Susquehanna River at four places on the image. Using an easily erasable pencil, draw the trend of this formation on the figure. Or, to put it another way, draw a continuous strike line along the crest of the ridge throughout its length. For the purpose of identification, this ridge-forming formation will be called formation A. It forms the flanks of plunging synclines where it is intersected by the Susquehanna River.

2. When you are satisfied that you have identified formation A by the method just described, draw over your pencil line with a yellow felt-tipped pen or pencil to distinguish it clearly from other ridges in the figure.

3. Locate the *next youngest* ridge-forming formation and trace its strike as you did for formation A. This younger formation will be called formation B, and it should be identified with a color on the image that contrasts with the one used for formation A.

4. Using a pencil, draw the axial traces of the folds that occur on the figure showing the direction of plunge and the symbol for an anticline or syncline. Reinforce your pencil line with a red pencil when you are certain of your interpretation.

5. Are the rocks west of the Susquehanna River (i.e., between the river and the right-hand margin of the figure) generally older or younger than those near the left-hand margin of the figure? Explain your reasoning.

Figure 4.16 Side-looking airborne radar (SLAR) image mosaic of part of the Harrisburg Map, Pennsylvania. Scale, 1:250,000. (U.S. Geologic Survey. Synthetic–Aperture Radar Imagery. Experimental Edition, 1982.)

North

Exercise 22. Interpretation of Geologic Maps

22A. LANCASTER GEOLOGIC MAP, Wisconsin

Six formations are shown on this map (fig. 4.17), each of which is identified by an abbreviation. The abbreviations and the names they represent are, in alphabetical order: Od, Decorah Formation; Og, Galena Dolomite; Op, Platteville Formation; Opc, Prairie du Chien Group; Osp, St. Peter Sandstone; and Qal, Alluvium. A *group* consists of two or more formations with significant features in common.

The contacts of these formations are more or less parallel to the topographic contours, thereby indicating that the formations are more or less horizontal. Another set of contours, shown in red, defines the top of the Platteville Formation (Op). The red numbers associated with these red contour lines indicate the elevation of the contour line above sea level.

1. Determine the oldest and youngest formations on the map and those of intermediate age. Complete the geologic column in figure 4.17 by printing the *abbreviation* of a formation in the appropriate box, and printing the *name* of the formation on the line immediately below the box.

2. Using the topographic contour lines, estimate the thickness of the Decorah Formation (Od), the Platteville Formation (Op), and the St. Peter Sandstone (Osp). Print the estimated thickness in feet of each of these formations to the right of the appropriate box in the geologic column. The thickness of a formation is determined by subtracting the elevation of the bottom of the formation from the elevation of the top of the formation. These elevations can be estimated from contour lines on either side of a contact.

3. Why is it impossible to determine the thickness of the Galena Dolomite and the Prairie du Chien Group?

4. Locate Cement School in the NE part of the map area. A road intersection near the school has an elevation of 1,076 feet and is so marked on the map. Using a nearby contour line showing the top of the Platteville Formation, determine how deep a well must be drilled at the road intersection to reach the top of the Platteville Formation.

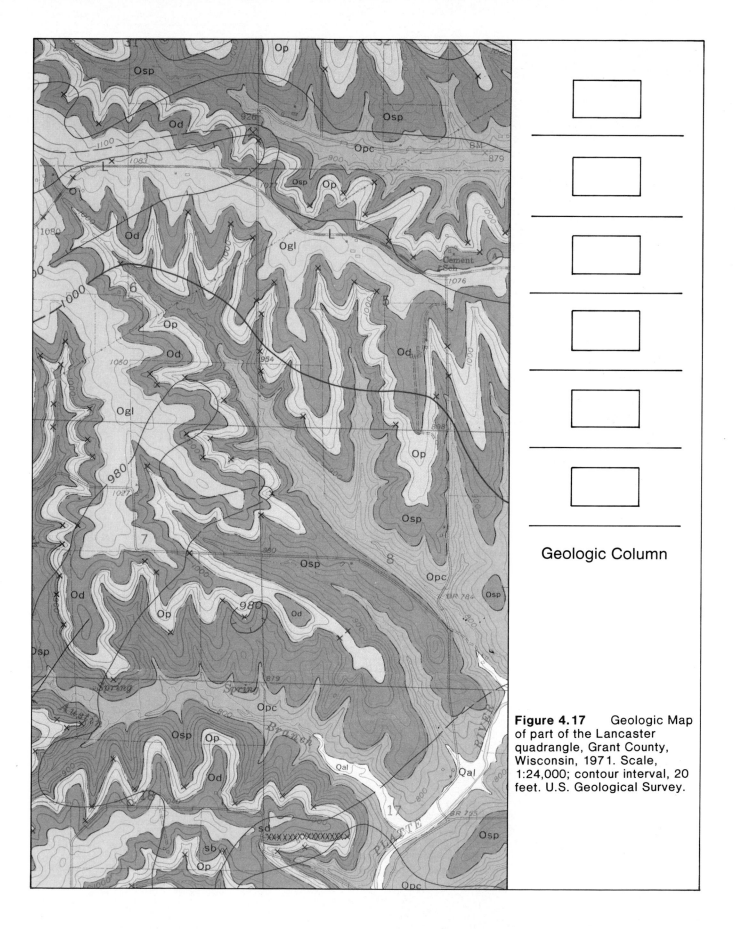

Geologic Column

Figure 4.17 Geologic Map of part of the Lancaster quadrangle, Grant County, Wisconsin, 1971. Scale, 1:24,000; contour interval, 20 feet. U.S. Geological Survey.

22B. SWAN ISLAND GEOLOGIC MAP, Tennessee

This geologic map (fig. 4.19) shows a banded outcrop pattern of sedimentary rocks. Study the map to get a general feel for the geology. Notice the strike and dip symbols on the map.*

1. The topographic profile of figure 4.18 is drawn along the black line on the map from the northwest to the southeast. Draw a geologic cross section along this line using figure 4.18 as the base. In aligning the profile of figure 4.18 with the line of profile on the map, place the point on the profile labeled Briar Fork at the point on the map where Briar Creek crosses the line of profile.

     The formations on the map are identified with letter symbols. To simplify the drawing of the geologic cross section, we will treat some of the formations as a *group* and show them as a group rather than as individual formations on the cross section. We will establish four arbitrary groups, each labeled by a letter symbol, and retain two single formations as they appear on the map. The groups and formations to be used in the cross section are as follows:

     Group Є: formations Єm, Єn, Єmn, and Єcr.
     Group LO: formations Oc, Ok, and Oma.
     Group MO: formations A, B, C, D, EFG, H, I, J, K, and LM.
     Formation Omb.
     Formation Os.
     Group SM: formations Sc and Mdc.

     On the cross section, label each group of formation according to the above scheme. Ignore formation Qal.

2. Complete the geologic column at the right of the map by placing the symbols for the four groups and two formations shown on your cross section in the six boxes provided. Use standard geologic practice, the <u>youngest at the top and the oldest at the bottom</u>.

*The number that appears next to a strike and dip symbol on a geologic map refers to the angle of dip of the rock layers as measured by a field geologist at a specific rock outcrop or exposure. When these numbers occur near a line on the map along which a geologic cross section is to be constructed, they should be considered approximations of the dip angles rather than absolute values. Dip angles within a few tens of feet of each other can vary as much as 5 to 10 degrees.

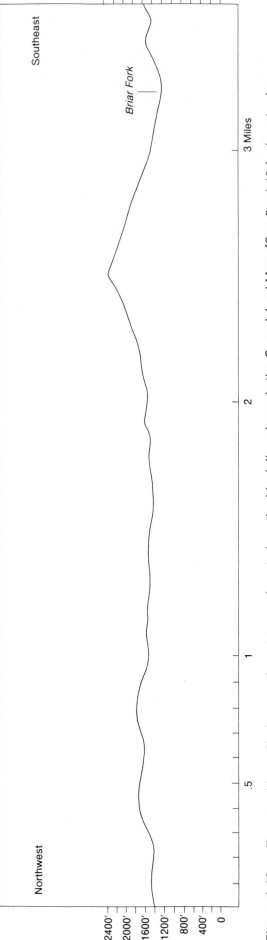

Figure 4.18 Topographic profile from northwest to southeast along the black line shown in the Swan Island Map. (See fig. 4.19 for location.)

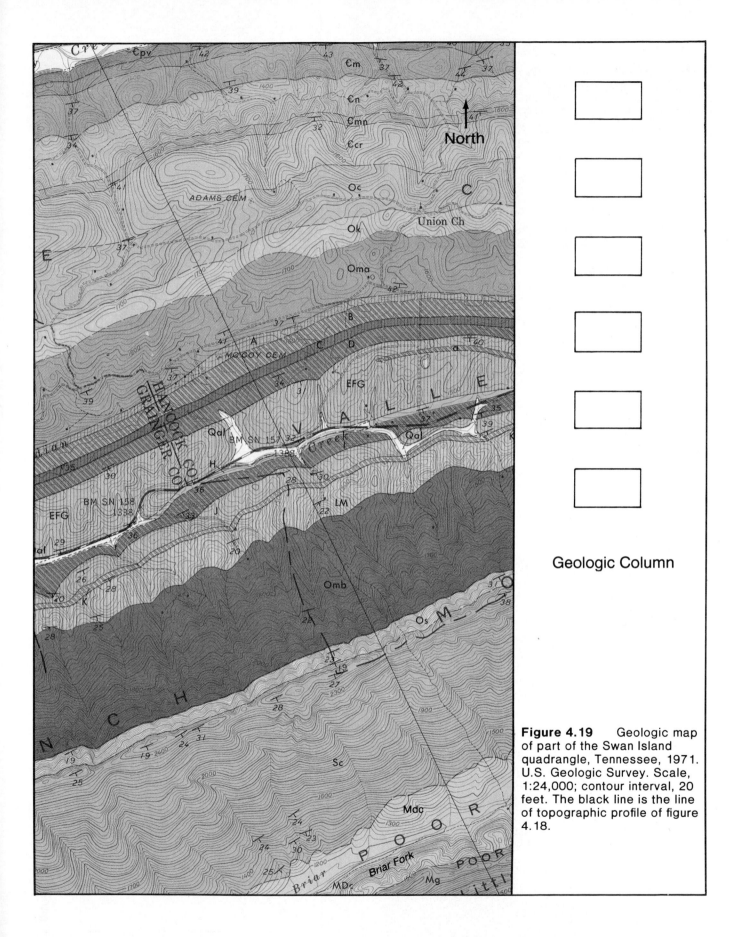

Geologic Column

Figure 4.19 Geologic map of part of the Swan Island quadrangle, Tennessee, 1971. U.S. Geologic Survey. Scale, 1:24,000; contour interval, 20 feet. The black line is the line of topographic profile of figure 4.18.

22C. COLEMAN GAP GEOLOGIC MAP, Tennessee-Virginia

This area (fig. 4.21) is underlain by sedimentary rock formations of different thicknesses. Note the many strike and dip symbols that occur on the map. (North is toward the bound edge of the page.)

1. The topographic profile of figure 4.20 is drawn along the line trending northwest-southeast across the map area from margin to margin, passing through the location of Brooks Well. Draw a geologic cross section along this line. Align the Brooks Well on the profile with the position of the Brooks Well on the map to achieve the proper correlation between the topography of the profile and the contours on the map. Note that the vertical scale of the profile is identical with the horizontal scale of the map. For assistance in drawing the geologic cross section, it should be noted that the Brooks Well penetrated the base of the Cc formation 300 feet below the ground surface. Color formations Cc and Ccr on the cross section, and label all formations with their correct symbols.

2. What is the thickness of formation Ccr?

3. Complete the geologic column to the right of the map. Put the symbol of the formation in the appropriate box and the name of the formation on the line below the box. The formational names and their respective symbols are as follows:
Cc, Conasauga Shale
Ccr, Copper Ridge Dolomite
Cmn, Maynardville Limestone
Ocl, Lower Chepultepec Dolomite

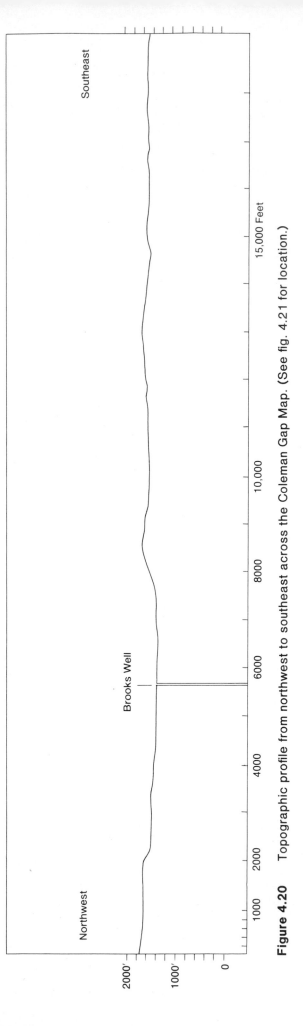

Figure 4.20 Topographic profile from northwest to southeast across the Coleman Gap Map. (See fig. 4.21 for location.)

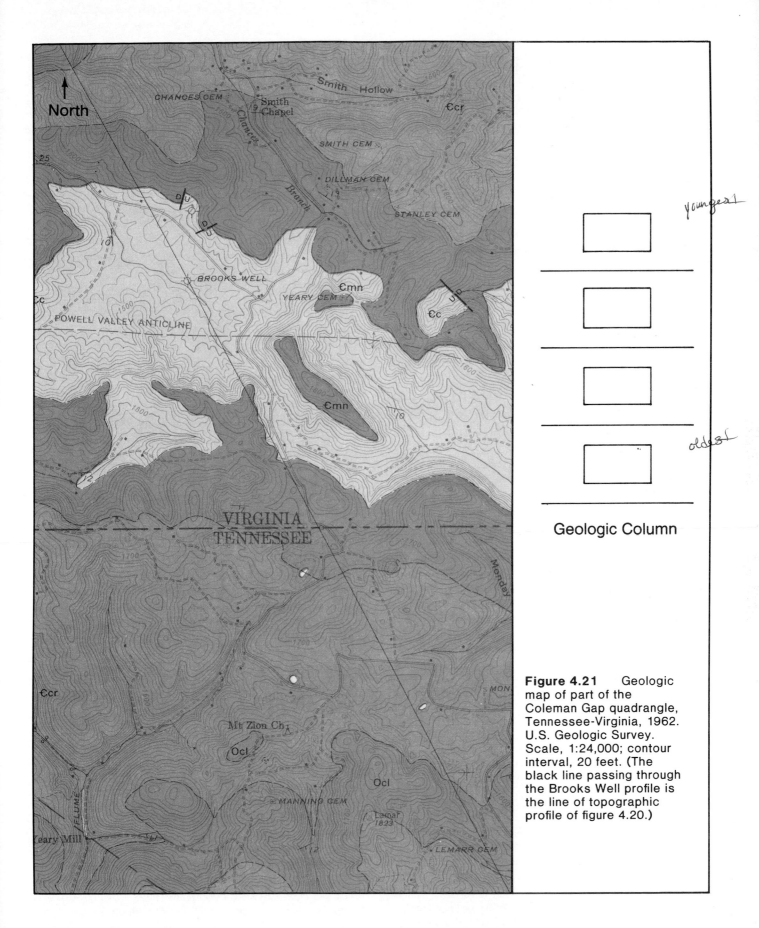

North

CHANCES CEM
Smith Hollow
Smith Chapel
€cr
25
SMITH CEM
DILLMAN CEM
STANLEY CEM
Branch
Chances
D/U
D/U
€mn
BROOKS WELL
YEARY CEM
€c
€c
POWELL VALLEY ANTICLINE
1500
1600
€mn
1600
10
1600
VIRGINIA
TENNESSEE
1700
1700
Monday
€cr
Mt Zion Ch
Ocl
Ocl
MON
MANNING CEM
FLUME
Lamar 1823
Yeary Mill
LEMARR CEM

youngest

oldest

Geologic Column

Figure 4.21 Geologic map of part of the Coleman Gap quadrangle, Tennessee-Virginia, 1962. U.S. Geologic Survey. Scale, 1:24,000; contour interval, 20 feet. (The black line passing through the Brooks Well profile is the line of topographic profile of figure 4.20.)

Faults and Earthquakes

Earth stresses that produce folds also produce faults. A *fault* is a fracture or break in the earth's crust along which differential movement of the rock masses has occurred. Movement along a fault causes dislocation of the rock masses on each side of the fault so that the contacts between formations are terminated abruptly.

Faults may be active or inactive. *Active faults* are those along which movement has occurred sporadically during historical time. Earthquakes are caused by movement along active faults. *Inactive faults* are those in which no movement has occurred during historical times. They are treated as part of the structural fabric of the earth's crust.

In this section we will deal first with inactive faults as part of structural geology, and secondly with active faults and their relationship to earthquakes.

Inactive Faults

A fault is a planar feature, and therefore its attitude can be described in much the same way that any geologic planar feature can be described. Figure 4.2 shows the various symbols used on a geologic map to define faults.

Nomenclature of Faults

Figure 4.22 is a block diagram of a hypothetical faulted segment of the earth's crust. The *fault plane* is defined as abcd. The fault plane strikes north-south and dips steeply to the east. A single horizontal sedimentary bed acts as a reference marker and shows that the displacement along the fault plane is equal to the distance x–y. This is called the *net slip*. The arrows show the direction of relative movement along the fault plane. Block A has moved up with respect to block B, and conversely, block B has moved down with respect to block A. Block A is called the *upthrown side* of the fault, and block B is the *downthrown side*. Block B is also known as the *hanging wall,* and block A as the *footwall.* Both terms are derived from miners who drove tunnels along fault planes to mine ore that had been emplaced there.

Faults generally disrupt the continuity or sequence of sedimentary strata, and they cause the dislocation of other

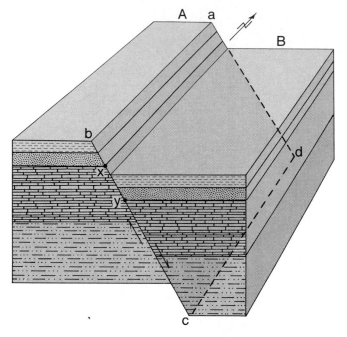

Figure 4.22 Block diagram of a fault. Arrows show the relative movement of block A with respect to block B. The horizontal beds have been dislocated a distance of x–y. The fault plane is abcd.

rock units from their prefaulted positions. On geologic maps, the intersection of the fault plane with the ground surface is called a *fault trace*. Fault traces are depicted on geologic maps by the use of standard symbols (fig. 4.2).

After faulting occurs, erosion usually destroys the surface evidence of the fault plane, so that with the passage of time the *fault scarp* (the exposed surface of the fault plane in figure 4.22) is destroyed. Only the fault trace remains.

Types of Faults

Faults are divided into three major types, each of which is defined by the relative displacement along the fault plane. In the first two types, the main element of displacement has been vertical, more or less parallel to the dip of the fault plane. A *normal fault* is one in which the hanging wall has moved down relative to the footwall (fig. 4.23*A*).

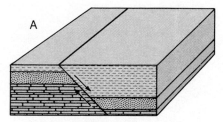

Normal Fault

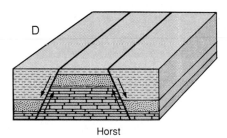

Reverse Fault

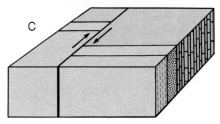

Strike-slip Fault

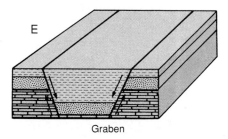

Horst

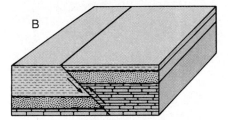

Graben

Figure 4.23 Block diagrams illustrating major fault types. Arrows indicate relative movement along the fault plane.

A *reverse fault* is one in which the hanging wall has moved up relative to the footwall (fig. 4.23*B*). A reverse fault in which the fault plane dips less than 45 degrees is called a *thrust fault*.

The third category of faults is characterized by relative displacement along the fault plane in a horizontal direction parallel to the strike of the fault plane. This type of fault is called a *strike-slip fault* (fig. 4.23*C*). Figure 4.23*D* shows a *horst,* an upthrown block bounded on its sides by normal faults. Figure 4.23*E* shows a *graben,* a downthrown block bounded on its sides by normal faults.

A fault shown on a geologic map can be analyzed to determine what kind of fault is involved. The analysis of normal and reverse faults will reveal the hanging and footwalls that lead to an understanding of the relative movement along the fault plane. In a strike-slip fault, offsetting of a marker bed as in figure 4.23*C* is the most direct evidence of the direction of movement. In some cases, the distinction between a strike-slip fault and a normal or reverse fault requires information not shown on the map.

A normal or reverse fault that cuts across the strike of inclined or folded sedimentary beds presents one of the most common situations for the analysis of movement along the fault plane. In such cases, there will be an apparent migration of the beds in the direction of dip of these beds on the upthrown side of the fault as erosion progresses. Stated another way, if an observer were to stand astride the fault trace, the observer's foot resting on the older rock would rest on the upthrown side. This is a simple mental test that can be applied to the analyses of faults presented in Exercise 23.

Exercise 23. Fault Problems

23A. FAULT PROBLEMS ON BLOCK DIAGRAMS

1. In figure 4.24, the upper group of three block diagrams shows (A) an unfaulted segment of the earth's crust with an incipient fault plane (dashed line), (B) movement along the fault plane, and (C) the appearance of the faulted area after erosion has reduced it to a relatively flat surface. The sequence D, E, and F is similar. Complete the outcrop patterns of the inclined formation on block diagrams C and F and label each as to the kind of fault (normal or reverse). Indicate by letters the hanging wall (H), the footwall (F), the upthrown (U), and the downthrown side (D) on each side of the fault trace in diagrams C and F.

2. Figure 4.25 shows geologic maps A and B, and their corresponding block diagrams. The sedimentary formations are numbered according to their ages (stratum 1 is the oldest). Complete the block diagram below each map and indicate by letters on the map: the hanging wall (H), footwall (F), upthrown side (U), downthrown side (D), and show by arrows the relative movement along the fault plane. Label each fault as either normal or reverse.

3. Figure 4.26 shows two separate sequences of three block diagrams: A, B, C, and E, F, G. A and E show conditions prior to faulting, B and F show conditions immediately after faulting, and C and G show the relationships on both sides of the fault traces after the fault scarp has been removed by erosion.

   a) Figure 4.26D is a geologic map of block diagram C. With a black pencil, draw the following symbols on the map: direction of dip of the fault plane; strike and dip symbols on the sedimentary beds on both sides of the fault trace; the upthrown (U) and downthrown (D) sides of the fault; and write either "normal" or "reverse" after the letter D at the lower margin of the map. (Refer to figure 4.2 to refresh your memory as to the various geologic map symbols.)

   b) In the blank space of figure 4.26H, draw a geologic map of block diagram G, draw the corresponding geologic map symbols as in 4.26D, and write either "normal" or "reverse" after the letter H on the lower margin of the map.

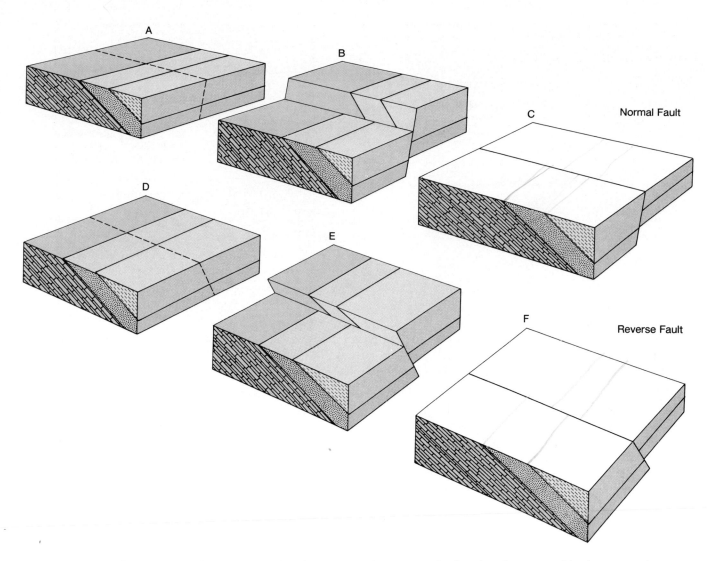

Figure 4.24 Series of block diagrams showing successive stages in the development of both a normal and a reverse fault. A and D before faulting with incipient fault plane marked by dashed line. B and E are shown immediately after faulting. C and F: after erosion of upthrown sides to a level common with the downthrown sides.

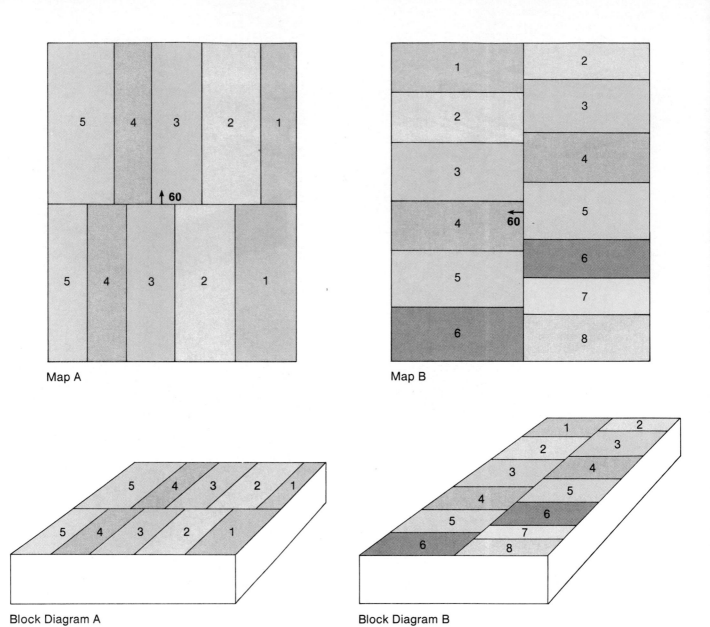

Map A

Map B

Block Diagram A

Block Diagram B

Figure 4.25   Block diagrams of faults and their corresponding geologic maps.

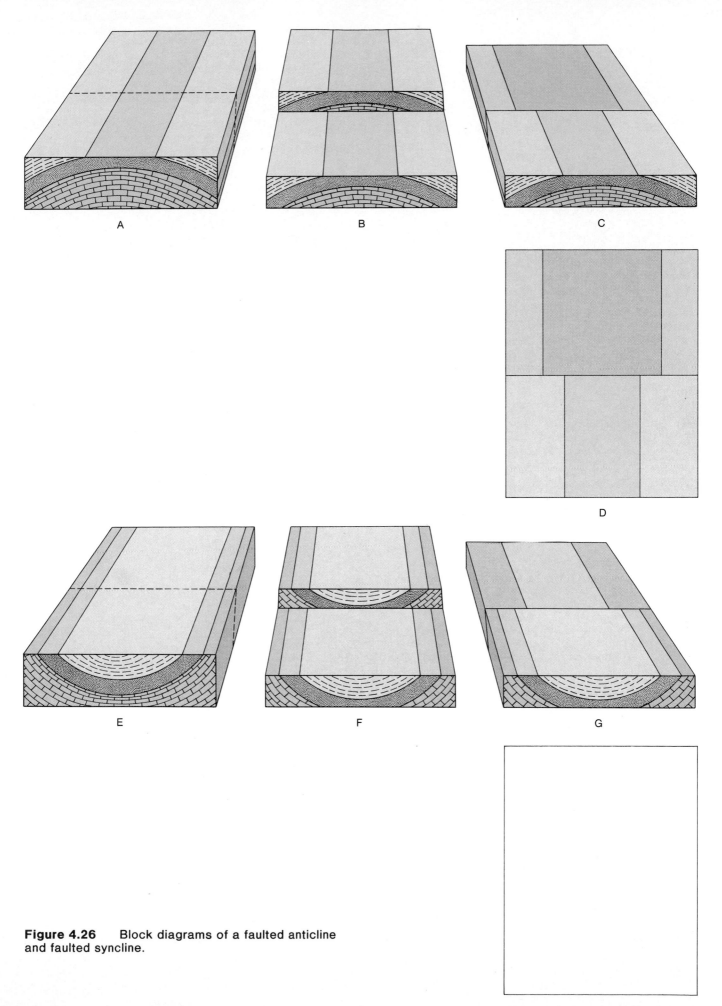

Figure 4.26 Block diagrams of a faulted anticline and faulted syncline.

23B. FAULTED SEDIMENTARY STRATA

1. On the Swan Island Map (fig. 4.27), formation Ok is cut by a fault trending NW-SE. Label the up-thrown and downthrown sides of the fault with the correct geologic symbols.

2. Refer to the fault described under question 1. What would be the direction of *dip* of the fault plane if this were a normal fault?

3. If the fault referred to in question 1 were a reverse fault, label the hanging-wall side with an H.

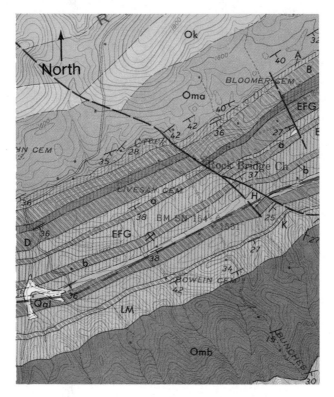

Figure 4.27 Part of the geologic map of Swan Island quadrangle, Tennessee, 1971, U.S. Geological Survey. Scale, 1:24,000; contour interval, 20 feet.

Active Faults

Fault Scarps and Fault Traces

The nomenclature of active faults is identical with the nomenclature of inactive faults. The attitude of an active fault plane may range from vertical to horizontal. Movement along the fault plane may produce a fault scarp at the earth's surface, but this scarp may be destroyed by erosion over time so that only a *trace* of the fault plane remains. Fault traces and fault scarps of both active and inactive faults can be identified on aerial photographs or images from earth-orbiting satellites by abrupt changes in topography or color patterns.

Earthquake Epicenters and Foci

The place where movement begins on the fault plane marks the *focus* of the resulting earthquake, and the point on the earth's surface vertically above the focus is the *epicenter* of the quake (fig. 4.28). The focus and epicenter of an earthquake that is generated on a vertical fault plane define the fault plane as a line on a cross section. If the earthquake is generated by movement along an inclined fault plane, the fault plane is defined in cross section by a line passing through the focus and the fault scarp or fault trace at the earth's surface. When movement occurs along a fault plane, energy is released and an earthquake is produced.

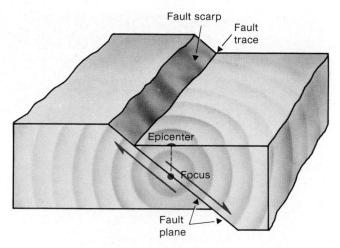

Figure 4.28   Block diagram showing a fault plane and the focus and epicenter of an earthquake generated by movement along the fault. Arrows show direction of relative movement along the fault and concentric shaded circles show the propagation of seismic waves. (From Carla W. Montgomery, *Physical Geology*, 2d ed. Copyright © 1990 Wm. C. Brown Publishers, Dubuque, Iowa. All Rights Reserved. Reprinted by permission.)

Exercise 24. Relationship of Fault Planes to Fault Traces, Epicenters, and Foci

The image of figure 4.29 extends from the Mohave Desert on the north to the Pacific Ocean on the south. Figure 4.30 is a generalized map of the same area showing the traces of *some* of the faults in the area. Those that are easily visible on the image are shown with a solid line, and those that are more difficult to recognize are shown by dashed lines.

1. Transfer the fault traces from figure 4.30 to figure 4.29.

2. Describe the physiographic features associated with the San Andreas and Garlock faults.

3. The San Andreas fault is a strike-slip fault that extends from the Gulf of California to beyond San Francisco in the Pacific Ocean, a distance of about 600 miles. The Pacific Ocean side of the fault has moved north (west in the area of the image) some 300 to 350 miles in a series of horizontal displacements. The San Francisco earthquake of 1906 was caused by slippage along the San Andreas fault in the amount of 21 feet.

    Figure 4.30 shows the epicenters of two major earthquakes in the greater Los Angeles area during historical times, the Ft. Tejon earthquake of 1857 and the Sylmar earthquake of 1971.

    a) Figure 4.31 is a schematic cross section of the earth across the San Andreas fault. The focus and the epicenter of the Ft. Tejon earthquake are shown. Draw a solid red line on the diagram showing the attitude of the San Andreas fault. What is the dip of the fault plane?

    b) Figure 4.32 is a schematic cross section through the focus of the 1971 Sylmar earthquake to the fault scarp caused by the Sylmar earthquake. Precise surveying after the Sylmar earthquake showed that the San Gabriel Mountains increased about 6 feet in elevation as a result of the movement along the San Fernando fault. This movement was responsible for the Sylmar earthquake.

    Draw a solid red line on figure 4.32 showing the San Fernando fault plane. Draw red arrows on each side of the fault indicating the relative movement along the fault plane. Label the hanging wall (H) and the footwall (F). Is the San Fernando fault a normal, reverse, or strike-slip fault? How determined?

Reference

Greensfelder, Roger. 1971. Seismologic and crustal movement investigations of the San Fernando earthquake. *California Geology,* April-May 1971: 62–68. California Division of Mines and Geology, Sacramento, California 95814.

Figure 4.29 False color image of the greater Los Angeles area of southern California, made from Landsat 1, October 21, 1972. (NASA ERTS image E-1090-18012.)

Structural Geology

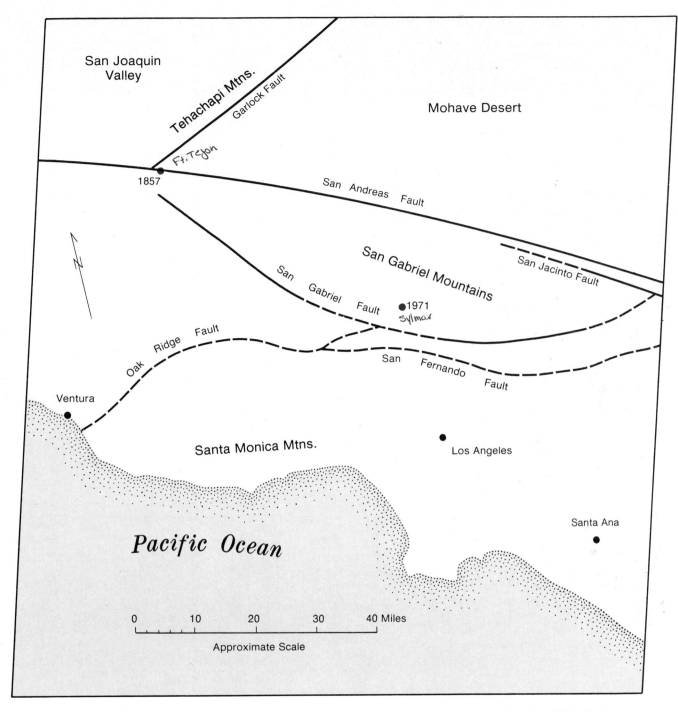

Figure 4.30 Generalized map of the greater Los Angeles area showing traces of some of the faults occurring there, also the epicenters of the Ft. Tejon earthquake of 1857 and the Sylmar earthquake of 1971. (R. H. Campbell, 1976. Active faults in the Los Angeles-Ventura area of southern California. *ERTS–1: A New Window on Our Planet, U.S.G.S. Professional Paper 929*, pp. 113–16.)

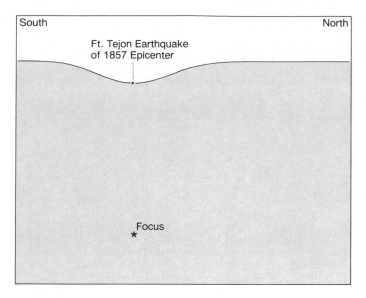

Figure 4.31 Schematic north-south cross section across the trace of the San Andreas fault showing the epicenter and focus of the Ft. Tejon earthquake of 1857.

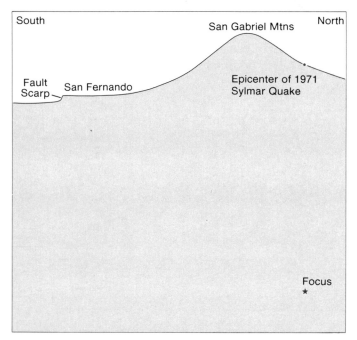

Figure 4.32 Schematic north-south cross-section from San Fernando across the San Gabriel Mountains through the epicenter of the Sylmar earthquake of 1971. Vertical scale exaggerated. (After R. Greensfelder, 1971. ''Seismologic and crustal movement investigations of the San Fernando earthquake.'' In *California Geology,* April–May 1971, p. 64.)

The Use of Seismic Waves to Locate the Epicenter of an Earthquake

Seismographs, Seismograms, and Seismic Observatories

The energy released by an earthquake produces vibrations in the form of *seismic waves* that are propagated in all directions from the focus. Seismic waves can be detected by an instrument called a *seismograph,* and the record produced by a seismograph is called a *seismogram.* The geographical location of a seismograph is called a *seismic station* or *seismic observatory,* and it is given a name in the form of a code consisting of three or four capital letters that are an abbreviation of the full name of the station. For example, a seismic observatory on Mt. Palomar, in southern California, has the code designation of PLM. A worldwide network of seismic observatories provides records of the times of arrival of the various kinds of seismic waves. Seismograms from at least three different stations located around the focus of a given earthquake but at some distance from it provide the data needed to locate the epicenter.

Seismic Waves

Two general types of seismic waves are generated by an earthquake: *body waves* and *surface waves.* Body waves travel from the focus in all directions through the earth; they penetrate the "body" of the earth. Surface waves travel along the surface of the earth and do not figure in the location of an epicenter.

Body waves consist of *primary waves* and *secondary waves.* The primary wave is referred to as the *P wave,* and the secondary wave is the *S wave.* The P wave is like a sound wave in that it vibrates in a direction parallel to its direction of propagation. An S wave, on the other hand, vibrates at right angles to the direction of wave propagation (fig. 4.33).

P and S waves are generated at the same time at the focus, but they travel at different speeds. The P wave travels almost twice as fast as the S wave and is always the first wave to arrive at the seismic station. The S wave follows some seconds or minutes after the first arrival of the P wave. *The difference in arrival times of the P and S waves is a function of the distance from the seismic station to the epicenter.* The distance from the seismic station to the epicenter is called the *epicentral distance.*

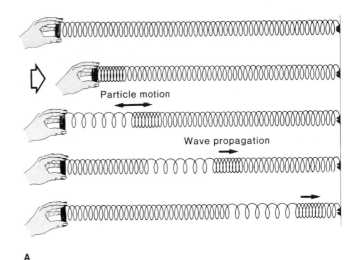

A

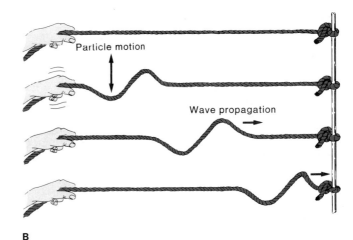

B

Figure 4.33 Particle motion in seismic waves. (*A*) A P wave is illustrated by a sudden push on the end of a stretched spring. The particles vibrate *parallel* to the direction of wave propagation. (*B*) An S wave is illustrated by shaking a loop along a stretched rope. The particles vibrate *perpendicular* to the direct wave propagation. (From Charles C. Plummer and David McGeary, *Physical Geology,* 4th ed. Copyright © 1988 Wm. C. Brown Publishers, Dubuque, Iowa. All Rights Reserved. Reprinted by permission.)

Reading a Seismogram

A seismograph records the incoming seismic waves as wiggly lines on a piece of paper wrapped around a drum rotating at a fixed rate of speed. The resulting seismogram contains not only the record of the incoming seismic waves but also marks that indicate each minute of time. Clocks at all seismic stations around the world are set at Greenwich Mean Time (GMT), so no matter what time zones observatories are located in, the seismograms produced at them are all based on a standardized clock.

When no seismic waves are arriving at an observatory, the seismograph draws a more or less straight line (fig. 4.34). Some small irregular wiggles on the seismogram may be *background noise* from vibrations produced by trucks, trains, heavy surf, construction equipment, and the like. Most modern seismographs contain a damping mechanism that reduces background noise to a minimum. In addition, background noise is kept to a minimum if the observatory is located in a remote area where human activities are uncommon.

The time of arrival of the first P wave is noted as T_p. The P wave continues to arrive until the first S wave appears, which is noted as T_s. The S wave has a much larger amplitude than the P wave. (The amplitude is the vertical distance between the peak of the recorded wave and the line on the seismogram recorded when no seismic waves are arriving.)

Figure 4.34 shows a seismogram from the Santa Ynez Peak Observatory on which the arrival times of the P and S waves are shown as T_p and T_s. These were determined by using the time scale on the seismogram to measure the time from the mark labeled 12:40:00 (12 hrs: 40 min: 00 sec) to the times of arrival of the first P and S waves. On the SYP seismogram of figure 4.34, T_p is 19 seconds after the time mark, or 12:40:19 GMT, and T_s is 44 seconds after the time mark, or 12:40:44 GMT. The difference between the time of arrivals, $T_s - T_p$, is therefore 25 seconds.

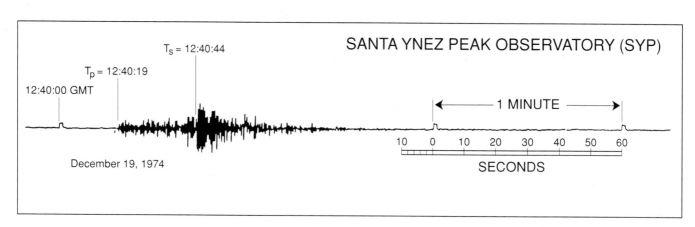

Figure 4.34 A seismogram recorded at the Santa Ynez Peak Observatory (SYP) in California showing an earthquake on December 19, 1974. The time mark automatically recorded on the seismogram is 12:40:00, which is 12:40 P.M. Greenwich Mean Time (GMT). The time of arrival of the first P wave, T_p, is 12:40:19, and the time of arrival of the first S wave, T_s, is 12:40:44. (Based on data from Charles G. Sammis, University of Southern California.)

Locating an Epicenter on a Map Using Travel-Time Curves

$T_s - T_p$ is measured in units of time, and this time, when converted to a distance, indicates the epicentral distance. Converting this time to distance requires the use of *travel-time curves* for both the P and S waves as shown in figure 4.35. A point on either one of the curves indicates the time required for a P or S wave to travel a certain distance from the epicenter. Time in seconds is shown on the vertical scale, and the corresponding distance in kilometers is shown on the horizontal scale. Following is the procedure for converting $T_s - T_p$ in seconds to an epicentral distance in kilometers:

1. Determine T_p and T_s from a seismogram to the nearest second. Record these values for use in the next step.
2. Subtract T_p from T_s and record as a time in seconds for use in the next step.
3. On the vertical axis of the travel-time graph of figure 4.35, set one point of a divider or compass on zero, and the other point on the value of $T_s - T_p$.
4. Move the compass upward and to the right until the point formerly on zero lies on the P curve and the other point lies on the S curve immediately above. It is important that the two points of the compass be on a vertical line in order to obtain the correct reading. Holding the compass in place, follow the vertical line on which the two points rest down to the horizontal scale, and read and record the epicentral distance.

As an example of this procedure, let us use the data from the SYP seismogram of figure 4.34. The value for $T_s - T_p$ on this seismogram is 25 seconds. With one point of the divider set on zero of the vertical axis of figure 4.35, we set the other point on 25. Then, with the compass at this setting, we move it between the P and S wave curves until one point of the compass is on the P wave curve and the other point is directly above it on the S wave curve. We follow the vertical line on which the two points rest down to the bottom scale and read 192 kilometers, the epicentral distance at station SYP.

5. On a suitable base map, use the bar scale on the map to *reset* your compass to the epicentral distance determined in step 4. Use this compass setting to draw a circle on the map whose center is at the geographic coordinates of the appropriate seismic station.
6. By following steps 1 through 5 for three different seismograms at appropriate directions and distances from the epicenter, you will draw three circles that intersect or nearly intersect at the epicenter.

Determining the Time of Origin of an Earthquake

The epicentral distances determined from $T_s - T_p$ are used to find the time of origin of the earthquake, designated by the symbol T_o. The procedure to do this is best explained by an example. Let us return to the information from the seismogram, recorded at station SYP, of figure 4.34. We have already determined that the epicentral distance is 192 kilometers. Remember that this is the distance from the seismic station to the epicenter of the earthquake. We want to know the time when this earthquake occurred, T_o. That is also the time when the seismic waves started their journey of 192 kilometers to SYP. Looking at figure 4.35, we see that the point where the P wave curve intersects the 192-kilometer line is 39 seconds. This tells us that it took the P wave 39 seconds to travel from the earthquake epicenter to station SYP. T_o is determined by subtracting the travel time of the P wave, 39 seconds, from T_p which is 12:40:19 GMT. Subtracting 39 seconds from 12 hrs, 40 minutes, 19 seconds gives us 12:39:40 GMT, the time of origin of the earthquake or T_o.

Exercise 25. Locating the Epicenter of an Earthquake

1. Table 4.1 gives the values of T_p and T_s from three seismograms that record an earthquake of December 19, 1974, in southern California. The seismic stations for each seismogram are shown on the map of figure 4.36 by their conventional abbreviations: SYP (Santa Ynez Peak), ISA (Lake Isabella), and PLM (Mt. Palomar). Use this map to plot the epicenter of the earthquake by using the information that you will enter in table 4.1. To systematize your work, fill in the blanks of table 4.1 before drawing the circles representing the epicentral distances. Follow the steps for locating an epicenter that are described earlier in this section. The data from the SYP Observatory, discussed above, appear as one of the three stations listed in

table 4.1. You can begin by filling in the blanks in table 4.1 with the epicentral distance and T_0 as already determined in the examples.

2. Which seismic station is closest to the epicenter? How is this determined?

3. Determine T_0 for the earthquake of December 19, 1974, using the data from stations PLM and ISA, and record it in table 4.1. Suggest a reason why the value for T_0 from each of the three stations is not the same.

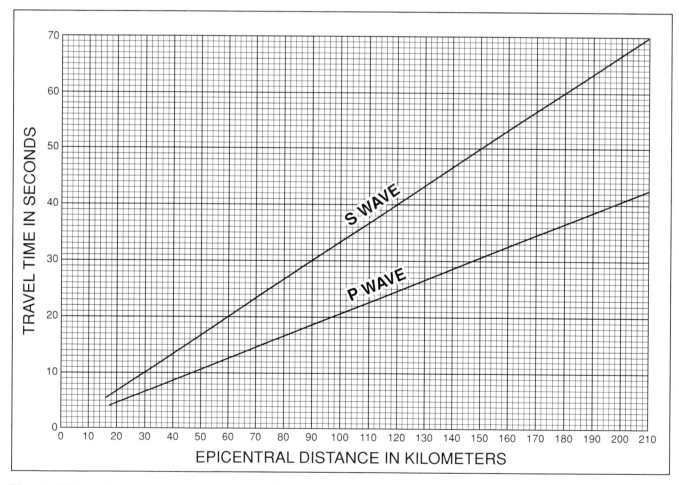

Figure 4.35 Travel-time curves for P and S waves in southern California. (Based on data from Charles G. Sammis, University of Southern California.)

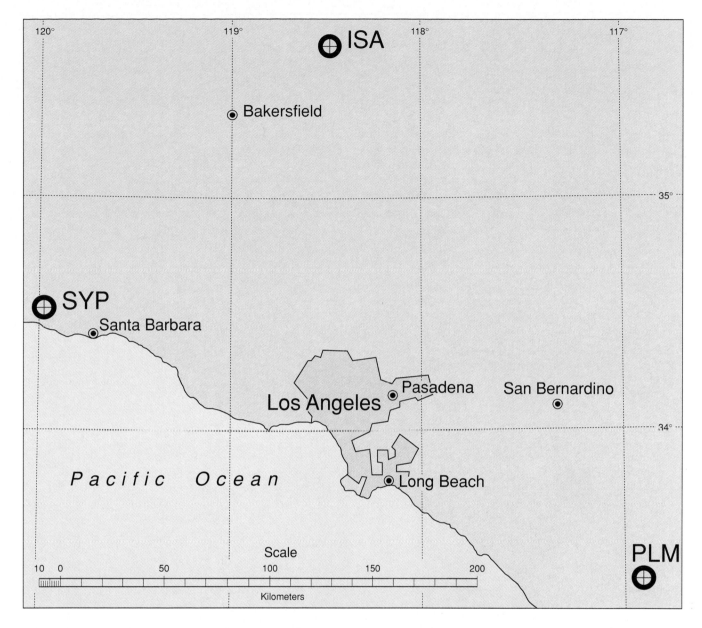

Figure 4.36 Base map of part of southern California showing location of seismic observatories: Santa Ynez Peak (SYP), Lake Isabella (ISA), and Mt. Palomar (PLM). (Lambert conformal conic projection, U.S. Geological Survey.)

Table 4.1 Arrival Times of P and S Waves at Seismic Stations Where They were Recorded During the Earthquake of December 19, 1974, in Southern California. To Be Filled In as Required in Exercise 25.

Station	Arrival Time, GMT		$T_s - T_p$, Seconds	Epicentral Distance, km	T_o
	P Wave (T_p)	S Wave (T_s)			
PLM	12:40:12	12:40:30			
ISA	12:40:18	12:40:39			
SYP*	12:40:19	12:40:44			

*Arrival times at this station are based on the seismogram in figure 4.34.

References

Eiby, George A. 1980. *Earthquakes.* Van Nostrand Reinhold Company, New York. 209 pp.

Bolt, Bruce. 1978. *Earthquakes: A primer.* W. H. Freeman & Company, San Francisco, Chapter 6.

We are indebted to Charles G. Sammis, Department of Geological Sciences at the University of Southern California, for his assistance in preparing this exercise.

Plate Tectonics and Related Geologic Phenomena

Part 5

Background

The theory of plate tectonics is a widely accepted concept which has been guiding geophysical and geological research since the mid-1960s. The term *plate tectonics* refers to the rigid plates that make up the skin of the earth and their movement with respect to one another. Figure 5.1 shows the distribution of the plates as they are currently understood. The plates differ greatly in size, although the true size of the plates is distorted in figure 5.1 because the map on which they are displayed is a Mercator projection. This kind of map represents the spherical shape of the earth on a flat surface and shows extreme distortion in the polar regions because the longitude lines on the map are parallel instead of converging (see fig. 2.4). It can be seen from figure 5.1 that some plates include both oceans and continents, as for example the Africa and South America plates.

Plate tectonics explains many phenomena on planet Earth, both on the continents and in the ocean basins, such as the origin and distribution of volcanoes and earthquakes, the topography of the sea floor, and a host of other major geologic features.

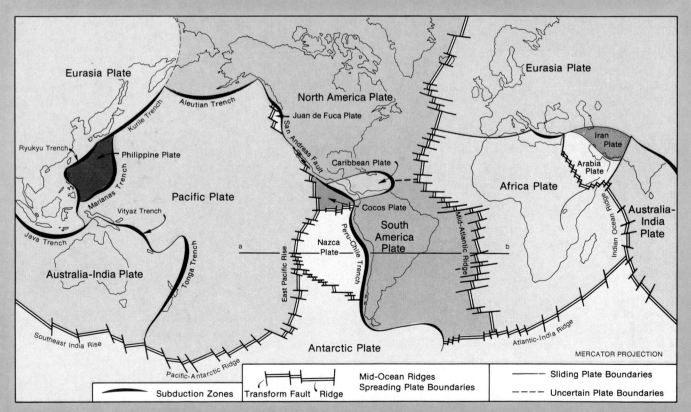

Figure 5.1   Map of the earth showing the names and boundaries of major lithospheric plates. The line labeled a–b is the line of the cross section of figure 5.2.

The Major Components of the Earth

The major components of the earth are the *lithosphere, mantle,* and *core*. The lithosphere is rigid and has two components, the *crust* and the *rigid upper mantle*. The *continental crust* is mainly granitic in composition, and the *oceanic crust* is mainly basaltic. The mantle consists of the rigid upper mantle, which along with the crust forms the lithosphere, and the *asthenosphere,* a weak, ductile layer on which the lithospheric plates move. (A ductile substance is one that deforms without fracturing.)

The relationship of these features is shown in figure 5.2, a scale diagram of a cross section of the earth from the surface to the core. The information shown on figure 5.2 has been gleaned mainly from the interpretation of seismic waves and laboratory experiments on rocks under high temperatures and pressures. None of the features in figure 5.2 have been observed directly in place, except, of course, the outermost crust at the earth's surface. Figure 5.2 is therefore an interpretation of the conditions prevailing at depth beneath the surface of the earth, based on the current theory of plate tectonics.

Plate Movement

The arrows in figure 5.2 show the direction of movement of the four plates covered in the cross section. This movement is deduced from the characteristics of the plate boundaries. Two plates that are moving away from each other form a *divergent* or *spreading boundary* along an *oceanic ridge* or *rise*. Basaltic magma is extruded at these boundaries to form new oceanic crust. Two plates that are moving toward each other form a *subduction zone,* where oceanic lithosphere descends beneath continental lithosphere along a deep ocean trench. Another kind of plate boundary is one where two plates slide past each other, as in the case of the San Andreas fault, which forms the boundary between the Pacific and the North America plates (fig. 5.1).

The forces that cause plate movement are not clearly understood, but the consensus is that *convection currents* are formed by differential heating of the mantle, which causes the asthenosphere to act like a giant conveyor belt along which the plates are rafted.

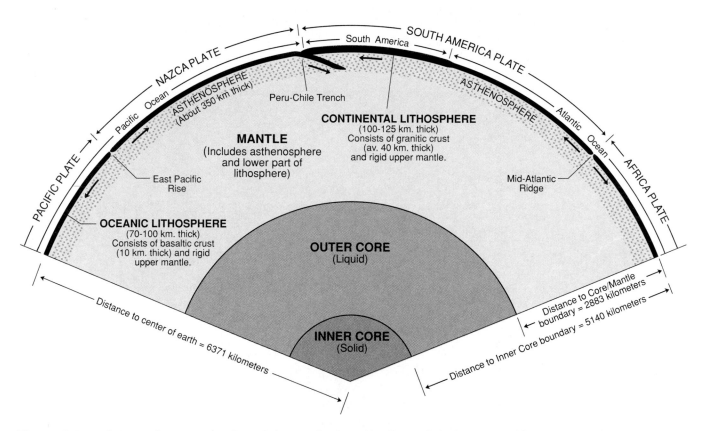

Figure 5.2 Schematic cross section of the earth along the line a–b in figure 5.1. The tectonic plates, the spreading plate boundaries coincident with the East Pacific Rise and the Mid-Atlantic Ridge, and the convergent plate boundary marked by the Peru-Chile Trench are shown. (The major parts of the earth's interior are based in part from Brian J. Skinner and Stephen C. Porter, 1987. *Physical Geology.* John Wiley & Sons, New York, figs. 2.2 and 2.3.)

Earth scientists believe that the surface area of the earth has remained constant over geologic time. Therefore, as new crustal material is formed along divergent boundaries by the extrusion of basalt from the underlying mantle, the creation of this new crust must be compensated by destruction of crust elsewhere. This is in fact what appears to happen in subduction zones where oceanic lithosphere slowly descends beneath adjacent continental lithosphere, as shown in figure 5.2 at the boundary between the Nazca and South America plates. The descending cool and rigid lithospheric slab is gradually incorporated into the hot mantle.

A larger scale schematic drawing of a hypothetical subduction zone is shown in figure 5.3. The oceanic lithosphere of basaltic crust and rigid upper mantle is subducted beneath the continental lithosphere. The foci of earthquakes in the subducted plate indicate that the subducted slab is still rigid; that is, it fractures or faults instead of flowing as is the case with the ductile asthenosphere. The earthquake foci in figure 5.3 are generalized to show that the shallow earthquakes are concentrated near the top of the descending plate and that the deeper earthquakes occur lower in the slab. This suggests that the upper part of the descending plate loses its rigidity as it moves deeper into the asthenosphere. The foci of earthquakes do not occur below about 700 kilometers. This fact supports the idea that the subducted slab loses its rigidity by the time it has reached that depth; it has literally been consumed in the mantle.

Earthquakes associated with divergent plate margins have shallow foci that are concentrated along the actively spreading ridge and the *transform faults* that offset the ridges. The pattern of these shallow earthquakes therefore roughly defines the location of the divergent plate margins.

Magma is generated above the subducted plate as the wet oceanic crust melts in contact with the hot asthenosphere, then rises to the surface where it is extruded as andesitic volcanoes.

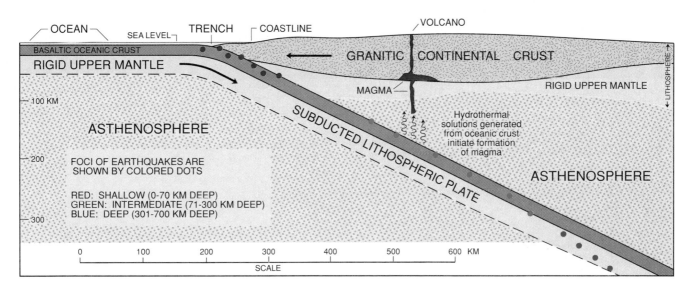

Figure 5.3 Schematic cross section of the earth showing the characteristics of a convergent plate boundary. The boundary is marked by a deep oceanic trench caused by the subduction of the oceanic lithosphere beneath continental lithosphere. The arrows show the direction of plate movement. The foci of earthquakes associated with the subducted plate do not all occur near the top of the plate. Those at greater depth are well below the top of the descending plate. (Based in part from M. Nafitoksöz, Nov. 1975. The subduction of the lithosphere, *Scientific American;* and Peter J. Wylie, et al., 1989. Interactions among magmas and rocks in subduction zone regions: experimental studies from slab to mantle crust, *Eur. Jour. Mineralogy,* v. 1, pp. 165–179.)

Exercise 26A. Plate Boundaries

Figure 5.4 is a Mercator projection on which the relief of the earth's surface features (general elevation above and below sea level) is shown by various colors. The range of elevations and depths of the oceans are shown in meters along the bottom margin of the map, and the highest and lowest points above sea level are shown in meters at the right and left ends of the color scale. The coastlines are shown by black lines that do not always coincide with the breaks in the color pattern.

The questions that follow, based on figures 5.1 and 5.4, are designed to make you familiar with the surface expression of the plate boundaries. Tear out figure 5.4 for ease of drawing on it and for referring to in this exercise.

1. Draw the boundaries of the tectonic plates shown on figure 5.1 on figure 5.4 with a pencil. When you are certain of the boundaries, reinforce your pencil line with a light-colored felt-tip pen so as not to obscure the detail on the map.

2. Which plate is *almost* totally surrounded by a spreading or divergent plate boundary?

3. Which plate is totally surrounded by a convergent boundary?

4. Which plates do not contain significant areas of continental land masses?

5. Name the plates that are bounded in part by the Mid-Atlantic Ridge.

6. Name the plates bounded in part by the East Pacific Rise.

7. What major island mass lies on the axis of the Mid-Atlantic Ridge?

Figure 5.4   Relief of the surface of the earth. The color bar across the lower margin of the map is the key to the range of elevations above sea level and the depths below sea level, in meters, represented by each color segment on the map. The coastline is shown by a black line which does not correspond with a color break on the map. (Source: Computerized Digital Image and Data Base available from the National Geophysical Data Center, National Oceanic and Atmospheric Administration, U.S. Department of Commerce, Code E/GC3, Boulder, Colorado 80303.)

Exercise 26B. The Nazca Plate

This exercise deals with the boundaries, or margins, of the Nazca plate (figs. 5.1 and 5.4) with special reference to its western boundary along the East Pacific Rise and its eastern boundary along the west coast of South America. You may wish to refer to figure 5.2 to review the general relationships of the Pacific plate and the South America plate. You will be drawing on figure 5.5 in completing this exercise, so before you begin, study figure 5.5 and notice particularly the distribution of shallow earthquake epicenters in the Pacific Ocean south of the equator, and the shallow, intermediate, and deep earthquake epicenters along the western side of South America.

1. Draw in black pencil on figure 5.5 the *approximate* boundary of the Nazca plate in the Pacific Ocean, using only the distribution of shallow earthquake epicenters (shown in red dots). Use figures 5.1 and 5.4 as a guide, but remember that only a rough boundary can be drawn from the earthquake data alone. Why do the shallow epicenters not fall on a smooth line?

2. Based on the distribution of earthquake epicenters alone, why is the boundary of the Nazca plate along the East Pacific Rise a divergent boundary rather than a convergent one?

3. What pronounced submarine feature along the eastern margin of the Nazca plate indicates that it is a convergent plate boundary?

(continued on p. 202)

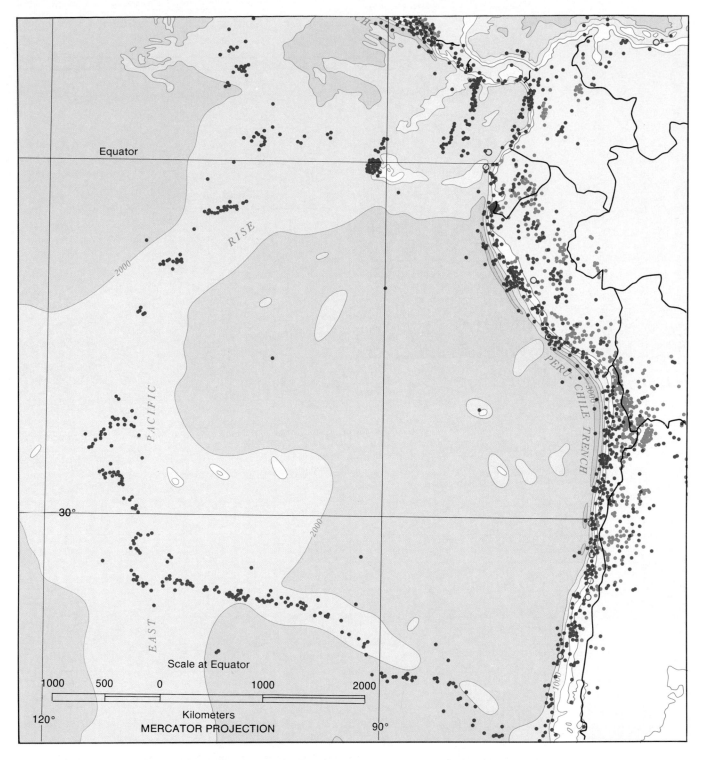

Figure 5.5 Map of part of the Nazca Plate. Epicenters of shallow earthquakes are shown in red; intermediate earthquakes in green; and deep earthquakes in blue. Ocean depths in meters. (Compiled from the *Plate Tectonic Map of the Circum-Pacific Region,* 1982; and the *Preliminary Tectonostratigraphic Terrane Map of the Circum-Pacific Region,* 1985. Maps copyrighted and published by the American Association of Petroleum Geologists, Post Office Box 979, Tulsa, Oklahoma 74101.)

4. Figure 5.6 is a generalized east-west profile from about the center of the Nazca plate to the eastern side of the Andes Mountains. The horizontal and vertical scales on figure 5.6 are the same. Ten epicenters of hypothetical earthquakes are shown on the profile, and the depths of their corresponding foci are listed in table 5.1. Plot the depth of each focus along the dashed line below each epicenter location, and mark the focus of each with a heavy dot in the appropriate color as used in figure 5.3 and 5.5.

5. Using the spatial pattern of earthquake foci that you have plotted on figure 5.6, construct a generalized cross section showing the tectonic features that exist below the profile. Use figure 5.3 for guidance and label the significant features.

Table 5.1 Depth to the Foci of the Earthquakes in Figure 5.6.

Earthquake Number	Depth to Focus (Kilometers)
1	15
2	25
3	30
4	50
5	150
6	180
7	200
8	225
9	350
10	375

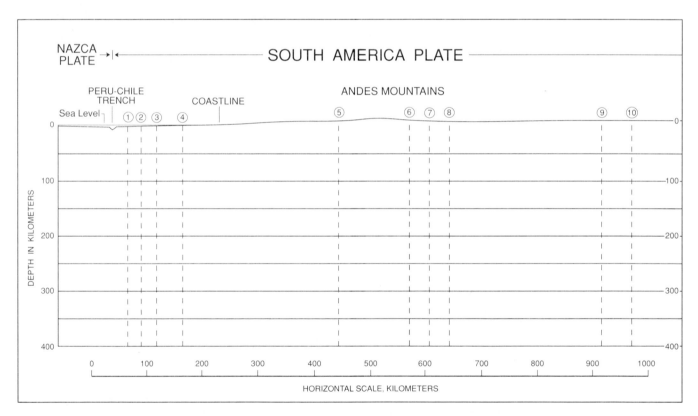

Figure 5.6 Cross section of part of the Nazca and South America Plates showing the epicenters of earthquakes identified in circles numbered 1 through 10. The depth to the focus of each earthquake is given in Table 15.1. This figure is to be used in answering questions 4 and 5 of Exercise 25B.

Seafloor Spreading in the South Atlantic and Eastern Pacific Oceans

Background

A basic premise of plate tectonics is that the crustal plates have moved with respect to each other over geologic time, and in fact are moving today. The rates of movement of crustal plates can be determined by using data from the plate margins along the mid-ocean ridges, where the amount of movement can be measured.

To measure the movement of two adjacent crustal plates along the margins of a divergent plate boundary, two things must be known: (1) two points on adjacent diverging plates that were once at the same geographic coordinates but have since moved away from each other over a known distance, and (2) the time required for the two points to move from their original coincident position to their present positions. If the two points can be identified and plotted on a map, the distance between them can be measured by use of the map scale. Determining the age in actual years of the two points involves the earth's magnetic field. It is therefore necessary to review this subject in order to understand how it relates to the movement of crustal plates.

The Earth's Magnetic Field

The earth is encompassed by a magnetic field. This field is analogous to the lines of force produced by a bar magnet with a north pole at one end and a south pole at the other. Imagine an immense bar magnet passing through the center of the earth with its north and south poles located near the north and south poles of the earth's axis of rotation (fig. 5.7A). A magnetic compass placed in this field would align itself parallel to the lines of magnetic force. The direction of this force is shown by the arrows in figure 5.7B, which point from the south magnetic pole toward the north magnetic pole. This condition is called *normal polarity*.

A magnetic compass does not point to the north geographic pole (true north) however, because the magnetic poles are not coincident with the geographic poles. The geographic poles define the earth's axis of rotation and remain fixed with respect to the equator over geologic time. The magnetic poles, however, shift over time with respect to the geographic poles. A plot of the magnetic poles during historical time shows that they tend to stay in close proximity to the geographic poles, so on a geological time scale, it is assumed that the magnetic poles and the geographic poles have remained within about 10 degrees of each other.

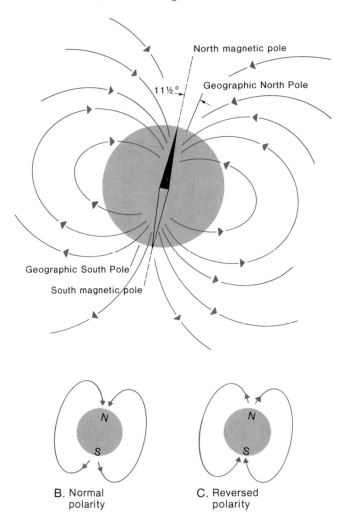

A. The earth's magnetic field.

B. Normal polarity

C. Reversed polarity

Figure 5.7 The earth's magnetic field. (*A*) The north magnetic pole and the north geographic pole are not coincident, and over geologic time they have been not much further apart than they are today. (*B*) Normal polarity characterizes the earth's magnetic field when the direction of the magnetic lines of force is from the south magnetic pole toward the north magnetic pole. (*C*) Reversed polarity occurs when the direction of the magnetic lines of force is toward the south magnetic pole. The earth's magnetic field has reversed many times during the history of the earth. (From Charles C. Plummer and David McGeary, *Physical Geology,* 4th ed. Copyright © 1988 Wm. C. Brown Publishers, Dubuque, Iowa. All Rights Reserved. Reprinted by permission.)

Reversals of the Magnetic Poles

A lava flow contains crystals of magnetite, a mineral with magnetic properties that aligns itself in the earth's magnetic field while the lava is still molten. When the lava solidifies, the magnetite minerals remain aligned parallel to the lines of force in the earth's magnetic field that existed when the lava cooled into rock. These magnetic crystals are like minute magnetic compasses frozen in the rock. Other rocks like sandstone also contain magnetic minerals that become aligned with the existing magnetic field when they sink to the lake bottom or sea floor at their site of deposition.

One of the amazing features of the earth's magnetic field is that its polarity has reversed itself many times over geologic time. That is to say that the north and south magnetic poles abruptly changed places so that a magnetic compass would point to the south magnetic pole during periods of *reversed polarity.*

The study of magnetism in ancient rocks is called *paleomagnetism.* The paleomagnetic features of rocks studied over the entire world have provided the basis for a detailed chronology of times of normal and reversed polarities during the last 170 million years. The rock layers from which this chronology has been assembled have been dated by radioactive means, thereby providing an absolute time scale that identifies the times when the periods of normal and reversed polarities occurred.

Both normal and reversed polarities are called *magnetic anomalies* or simply *anomalies.* A chronology of magnetic anomalies is given in figure 5.8. Three elements are contained in this chronology: (1) a time scale in millions of years before the present (Ma Age in figure 5.8), (2) the periods of normal (black) and reversed (white) polarities, and (3) the conventional identification numbers that have been assigned to each anomaly. These identification numbers are arranged in chronological order with the youngest anomaly designated by number 1, and successively older anomalies by numbers 2, 2a, 3, and on up to anomaly 33. (Figure 5.8 is a shortened version of a chronology that extends to anomaly 37, which is about 170 million years old.)

The time scale of figure 5.8 will be used later in an exercise, so it will be useful to become familiar with it. For example, the anomaly with the identification number 6 is a positive anomaly that is 20 million years old. Notice that the identification number assigned to an anomaly bears no relationship to that anomaly's absolute age. In other words, the series of identification numbers represents a *relative chronology,* and the series of dates in millions of years is an *absolute chronology.*

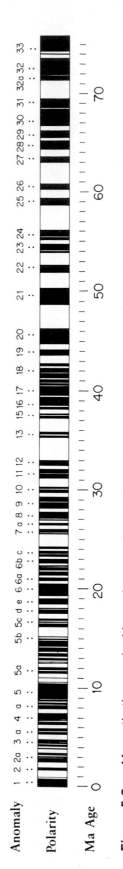

Figure 5.8 Magnetic time scale. Magnetic anomaly identification numbers are given on the left-hand side and an absolute time scale in millions of years (Ma) is given on the right-hand side. Normal magnetic anomalies are shown in black, and reversed polarity is shown in white. The absolute age of the anomalies and their polarities can be read directly from the chart. For example, anomaly 4 is 7 million years old, and it is a positive anomaly. (From *Magnetic Lineations of the World's Ocean Basins.* Copyright © 1985 the American Association of Petroleum Geologists, Post Office Box 979, Tulsa, Oklahoma 74101. All Rights Reserved. Reprinted by permission.)

Magnetic Anomalies

An anomaly is a departure from the normal scheme of things. With respect to rocks that contain magnetic minerals, a *magnetic anomaly* is a magnetic reading that is greater or less than the normal strength of the magnetic field where the rock occurs. To illustrate, consider a series of bands of lava beds (basalt) lying on the sea floor along both sides of an active spreading ridge between two crustal plates. An instrument that measures the intensity of the magnetic field is called *magnetometer*. When one is towed on a long cable behind a ship headed along a course across the trend of the lava beds, the strength of the earth's magnetic field is recorded continuously on board the ship, and a satellite navigational system simultaneously records the ship's position.

When the seaborne magnetometer passes over basalts that are normally polarized, the strength of the earth's magnetic field is slightly intensified because the ancient magnetism in the normally polarized rocks adds a small component to the normally polarized earth's field. In this case, a *positive anomaly* is recorded. If the rocks over which the magnetometer is passing were formed during a time of reversed polarity, the strength of the earth's field is slightly reduced and a *negative anomaly* is recorded.

Figure 5.9*A* shows a hypothetical record of magnetic field strengths in relation to basalts on either side of a mid-ocean ridge. High points on the curved line indicate a normal or positive anomaly, and low points indicate a reversed or negative anomaly. The pattern of negative and positive anomalies is repeated on either side of the spreading ridge; that is, the magnetic curve on one side of the ridge is a mirror image of the curve on the opposite side. The spreading ridge crest is thus flanked by stripes of alternating positive and negative anomalies *that increase in age with distance from each side of the spreading ridge* (fig. 5.9*B*).

One can deduce from this relationship that the tectonic plates are moving away from the divergent plate boundary. Moreover, if the *absolute age* of the various anomalies on either side of the ridge is known, and the distance between anomalies of the same age is measured, the *spreading rate* of the two adjacent plates can be determined. Using such information, the previous relative positions of two continents, such as South America and Africa, can be determined for various times in the geologic past.

In Exercise 27 you will determine the spreading rates on segments of the East Pacific Rise and the Mid-Atlantic Ridge, and in Exercise 28 you will determine the relative positions of South America and Africa at a specific time in the geologic past.

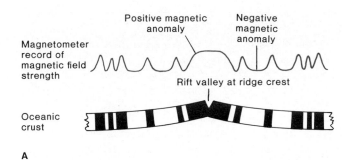

Figure 5.9   Marine magnetic anomalies. (*A*) The red line shows positive and negative magnetic anomalies as recorded by a magnetometer towed behind a ship. In the cross section of oceanic crust, positive anomalies are drawn as black bars and negative anomalies are drawn as white bars. (*B*) Perspective view of magnetic anomalies shows that they are parallel to the rift valley and symmetric about the ridge crest. (From Charles C. Plummer and David McGeary, *Physical Geology*, 4th ed. Copyright © 1988 Wm. C. Brown Publishers, Dubuque, Iowa. All Rights Reserved. Reprinted by permission.)

Exercise 27. Spreading Rates on the East Pacific Rise and the Mid-Atlantic Ridge

1. Figure 5.10 is a map of the magnetic anomalies (also called magnetic lineations) between the Gofar Fracture Zone (F. Z.) and the 23 Degree South Fracture Zone along the East Pacific Rise. The active spreading ridge corresponds to the plate margin between the Pacific plate and the Nazca plate. The active spreading ridge is shown as black bands offset by fracture zones, and the numbered lines on either side of the spreading ridge are magnetic anomalies or magnetic lineations. The age of each numbered anomaly can be determined from figure 5.8. Record the ages of anomalies 2, 3, 5, and 5a in table 5.2 in the second column.

2. Measure the distance between each pair of anomalies of the same age on opposite sides of the spreading ridge along the Garrett Fracture Zone, using the bar scale on the map. Record these distances opposite the appropriate anomaly in table 5.2.

3. Plot a point on the grid of figure 5.12 for each anomaly listed in table 5.2. Each point will be located where the age in millions of years along the vertical axis intersects the distance in kilometers along the horizontal axis.

4. Draw a straight line that best fits the points plotted on the grid of figure 5.12. The line must begin at the intersection of the horizontal and vertical axes of the grid. Label this line "East Pacific Rise."

5. Measure the distances between the magnetic lineations on the Mid-Atlantic Ridge shown on figure 5.11. Make the measurements along the Ascension Fracture Zone. (Note: anomaly 13 is shown only on the east side of the ridge. To make this measurement compatible with the others, measure the distance from anomaly 13 to the center of the Mid-Atlantic Ridge and multiply by 2.) Record the distances and ages in the appropriate columns of table 5.2.

6. On figure 5.12, plot the age vs. distance for each point, draw a best fit straight line through the points, and label the line "Mid-Atlantic Ridge."

7. The two straight lines on figure 5.12 provide a visual comparison between the spreading rates of the East Pacific Rise and the Mid-Atlantic Ridge. Is the spreading rate greater on the East Pacific Rise or the Mid-Atlantic Ridge?

8. Calculate the spreading rate in centimeters per year (cm/yr) using information based on anomaly number 5 on the East Pacific Rise, and anomaly number 5 on the Mid-Atlantic Ridge. Remember that 1 km = 1,000 m, and 1 m = 100 cm.

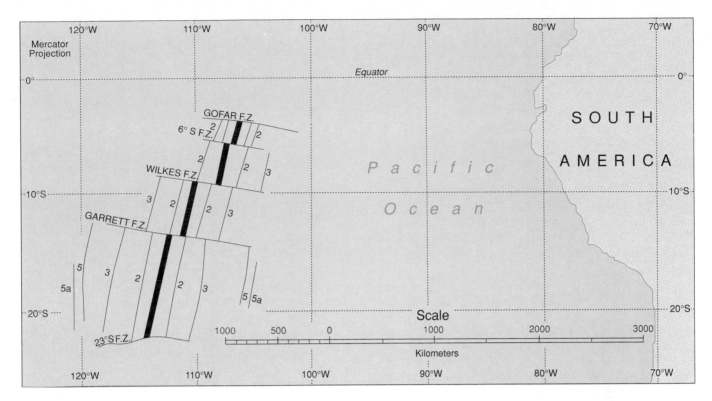

Figure 5.10 Map of part of the Pacific Ocean showing a segment of the East Pacific Rise. The active spreading ridge is shown by black bars. The active spreading ridge has been offset by the fracture zones (F.Z.). Ages of the numbered magnetic anomalies or magnetic lineations can be determined by reference to the magnetic time scale in figure 5.8. (From *Magnetic Lineations of the World's Ocean Basins.* Copyright © 1985 American Association of Petroleum Geologists, Post Office Box 979, Tulsa, Oklahoma 74101. All Rights Reserved. Reprinted by permission.)

Table 5.2 Ages of Selected Magnetic Anomalies on the East Pacific Rise and the Mid-Altantic Ridge and the Distances Between Them.

Anomaly Number	Age in Millions of Years	Distance Between Anomalies in Km	
		East Pacific Rise	Mid-Atlantic Ridge
2			/////
3			/////
5			
5a			/////
8		/////	
13		/////	
21		/////	

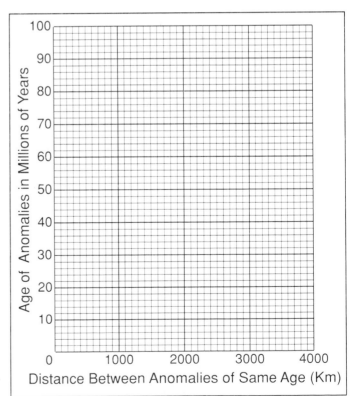

Figure 5.12 Grid on which to plot the ages of anomalies against the distances separating them across the East Pacific Rise and the Mid-Atlantic ⟶ Ridge. For use in answering questions in Exercise 27.

Exercise 28. Restoration of the South Atlantic Coastline 50 Million Years before the Present

Given the evidence of spreading along the Mid-Atlantic Ridge, it can be deduced that the Africa and South America *plates* are moving away from each other, carrying the *continents* of Africa and South America with them. By using the pattern of magnetic lineations shown on figure 5.11, it is possible to reverse the spreading process and restore the positions of the African and South American coastlines to a time when a particular set of magnetic lineations was being formed on the Mid-Atlantic Ridge.

For the purposes of this exercise we will use anomaly number 21, which according to the magnetic lineation time scale of figure 5.8 was formed 49.6 or roughly 50 million years ago. Proceed as follows.

1. On figure 5.11, draw a red line over each of the magnetic lineations of anomaly number 21 on the South American side of the Mid-Atlantic Ridge. Connect the segments of the number 21 anomaly with a red line drawn along the fracture zones against which they terminate. Start with the point where anomaly 21 touches the Ascension F. Z. Follow anomaly 21 with your red pencil southward until it reaches the Bode Verde F. Z., then along the Bode Verde F. Z. westward to the northern end of the next fracture zone. Continue until you have reached the southernmost fracture zone on the map.

2. Attach a piece of tracing paper over figure 5.11 with tape or paper clips, and repeat the process described above for anomaly 21 on the African side of the Mid-Atlantic Ridge. Draw this line in red pencil on the tracing paper.

3. With the tracing paper still in place, trace the coastlines of Africa and South America on the tracing paper with black pencil. Also, trace on the tracing paper the boundaries of figure 5.11 and the 20° South latitude line in black pencil.

4. Detach the tracing paper and slide it toward South America until the red line on the tracing paper matches the red line on figure 5.11. When the two lines are matched as closely as possible, hold the tracing paper in place and trace the coastline of South America in red pencil on the tracing paper. Trace also the 20° South line on the tracing paper in red pencil.

5. The map you have constructed on the tracing paper shows the Mid-Atlantic Ridge as it existed when magnetic anomaly 21 was being formed. Your tracing paper also shows the *relative* positions of segments of the coastlines of Africa in black pencil and South America in red pencil as they were approximately 50 million years ago. This reconstruction is based on the assumption that the *continents* of Africa and South America were fixed to their respective plates during the spreading process over the past 50 million years. The *continents* moved with respect to each other because the *tectonic plates* to which they were attached moved as spreading continued along the Mid-Atlantic Ridge. What is the evidence that the movement of the two plates was not strictly in an east-west direction?

6. Was the earth's magnetic field normal or reversed at the time represented by your map on the tracing paper?

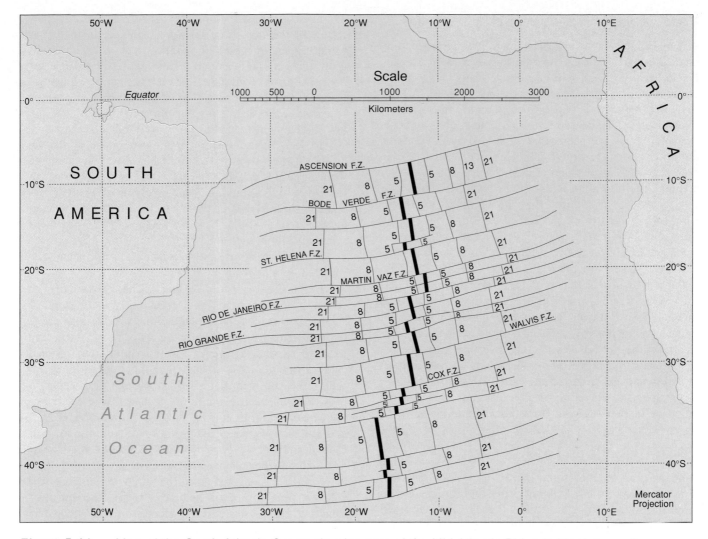

Figure 5.11 Map of the South Atlantic Ocean showing part of the Mid-Atlantic Ridge in black bars, the east-west fracture zones, and selected magnetic anomalies. The ages of the numbered anomalies or magnetic lineations can be determined from the magnetic time scale in figure 5.8. (From *Magnetic Lineations of the World's Ocean Basins.* Copyright © 1985 American Association of Petroleum Geologists, Post Office Box 979, Tulsa, Oklahoma 74101. All Rights Reserved. Reprinted by permission.)

Volcanic Islands and Hot Spots

The Hawaiian Islands consist of a northwest-trending chain of volcanic islands. Only the largest of these, Hawaii, has active volcanoes, whose eruptive history you are already familiar with from a previous exercise in this manual. The other islands in the chain are also volcanic in origin, but they ceased erupting some millions of years ago and have since been severely eroded by wave action and intense surface runoff. The Hawaiian Islands are, in fact, part of a longer chain of extinct volcanic islands that compose the Hawaiian Ridge (fig. 5.4).

It has been postulated that the alignment of the extinct volcanoes forming the Hawaiian Ridge was formed as follows. A *hot spot,* whose latitude and longitude has remained fixed over many millions of years, lies in the asthenosphere beneath the island of Hawaii. This hot spot is the source of heat that produces the magma that feeds the volcanoes on Hawaii. The oceanic lithosphere has been moving northwest over this spot in a more or less straight line, and as each nonvolcanic part of the ocean floor is carried over the hot spot by the conveyor action of the asthenosphere, a new volcano is born. Evidence in support of this hypothesis lies in the absolute dates of the old lava flows along the Hawaiian Ridge. These are successively older the farther they are from the island of Hawaii.

Volcanic Islands and Atolls

An *atoll* is an oceanic island that in map view appears as a narrow strip of land with low relief that forms a closed loop. Inside the loop is a shallow lagoon. The loop itself may contain gaps that allow access by ships from the surrounding ocean to the lagoon. An atoll is made chiefly of *coral reef,* but it also contains assorted other marine organisms such as algae and mollusks. Some of the coral may be weathered and eroded by wave action to form *coral sand.*

Coral is a marine animal that thrives in warm shallow waters of the world's oceans, most notably in the equatorial regions of the Pacific Ocean. Reef-forming corals grow in colonies that attach themselves to the shallow sea floor in subtropical and tropical climatic zones. Corals live in water that is no more than about 50 meters deep, is relatively free of sediments, is penetrated by sunlight, and has an abundant food supply of small marine organisms.

In 1842, Charles Darwin proposed a theory for the origin of atolls that he based on his observations of the tropical islands during the voyage of the *Beagle* through the equatorial waters of French Polynesia. Darwin recognized three stages in the evolution of atolls in the islands around Tahiti. These stages are illustrated schematically in figure 5.14.

Stage I consists of a newly formed volcanic island surrounded by a *fringing coral reef.*

Stage II is reached after the volcanic peak has been eroded by surface runoff and wave action, and has partly subsided beneath the sea. As the island subsides, the corals of the fringing reef die because the water becomes too deep for their survival. However, the remaining mass of dead coral forms a platform on which new corals establish themselves continuously. This process allows the upward growth of the reef to maintain pace with the sinking of the island. The fringing reef now becomes a *barrier reef,* and a shallow lagoon develops between it and the shore of the volcanic island. The low-lying surface of the barrier reef also may be colonized by vegetation.

Stage III is reached when the eroded remnants of the volcanic peak disappear beneath the sea through continued surface erosion and subsidence of the sea floor. The coral reefs continue their upward growth and ultimately encircle the enclosed lagoon.

Darwin's theory was challenged by others on the grounds that no mechanism was known that could cause a volcanic island in the middle of the ocean to sink. These

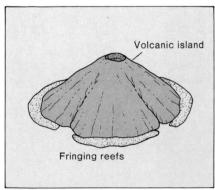
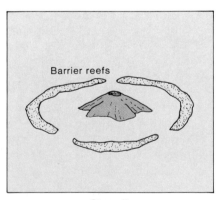

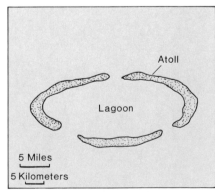

Figure 5.14 Diagrams of the three stages in the evolution of an atoll as proposed by Charles Darwin. See text for description of each stage. (From Charles C. Plummer and David McGeary, *Physical Geology,* 4th ed. Copyright © 1988 Wm. C. Brown Publishers, Dubuque, Iowa. All Rights Reserved. Reprinted by permission.)

critics proposed instead that the evolution of an atoll was due to a rising sea level, not a sinking island. With the advent of plate tectonics, however, Darwin has been vindicated. Volcanic islands are formed by submarine eruptions over hot spots and are carried on the oceanic lithosphere to deeper water by plate movement.

The rising sea level proponents of atoll evolution were not totally wrong, however. Sea level has fallen and risen during past geologic time in response to the waxing and waning of Pleistocene ice sheets, and these sea-level fluctuations account for some *dead* coral reefs that now lie *above* sea level on the flanks of some volcanic islands.

The modern theory of atoll formation, therefore, calls for a subsiding volcanic island on which sea-level fluctuations over geologic time have been superimposed.

Exercise 29A. Movement of the Volcanoes in the Hawaiian Ridge Over the Hawaiian Hot Spot

1. The map of figure 5.13 shows part of the Hawaiian Ridge and the absolute dates of lava along it printed in bold black numbers that represent millions of years before the present. Hawaii contains an active volcano, Mauna Loa, so the lava from it is zero years old. The lava on Nihoa Island is 7 million years old. Thus, according to the hot spot hypothesis, Nihoa Island was once an active volcano standing where Hawaii stands today.

   a) Determine by simple arithmetic the rate of movement of the oceanic lithosphere over the Hawaiian hot spot. Figure the rate in centimeters per year using the distances from Hawaii to each of the three dated lavas on the map of figure 5.13. Make your distance measurements from the center of the zero on Hawaii to the center of each of the boldface numbers on the ridge.

   b) Do the rates of movement based on the three dates indicate that the movement has been constant or variable?

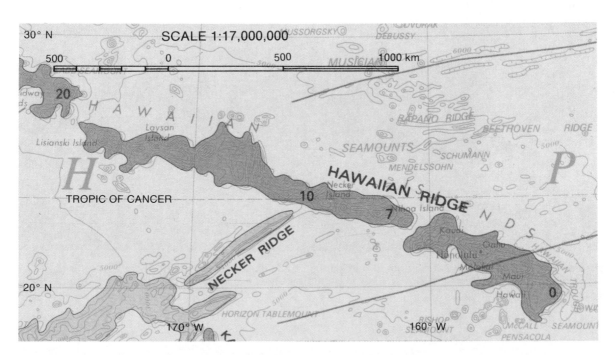

Figure 5.13 Map of the Hawaiian Ridge in the Pacific Ocean. Contour intervals, 1,000 meters. (From the *Preliminary Tectonostratigraphic Map of the Circum-Pacific Region.* Copyright © 1985 the American Association of Petroleum Geologists, Post Office Box 979, Tulsa, Oklahoma 74101.)

Exercise 29B. Islands in French Polynesia of the South Pacific Ocean

1. Figures 5.15, 5.16, and 5.17 represent the three stages in atoll evolution as proposed by Darwin. What are the main identifying characteristics of each that can be observed in the photographs?

2. What is the cause of the gaps in the fringing reef of figure 5.15?

3. What kind of rock would you expect to be encountered by a drill that penetrated the coral and sediments in the center of the lagoon in figure 5.17?

4. Assume that another hole was drilled somewhere on the atoll itself in figure 5.17.

a) Would the thickness of the coral penetrated by the drill be greater or less than in the center of the lagoon?

b) Sketch a cross section through the center of the atoll in figure 5.17. Exaggerate the vertical scale and show the following:
   1) The foundered volcanic island remnant.
   2) Sea level.
   3) Coral reef material above and below sea level.
   4) The location of the two drill holes: A, through the lagoon; B, through the atoll itself.

Figure 5.15   Moorea Island in the Society Islands of French Polynesia in the equatorial Pacific Ocean, an example of Stage I in the evolution of an atoll. (With the permission of and copyrighted by Erwin Christian.)

Figure 5.16   Bora-Bora Island in the Society Islands of French Polynesia in the equatorial Pacific Ocean, an example of Stage II in the evolution of an atoll. (With permission of and copyrighted by Erwin Christian.)

Figure 5.17   Aratica Island in the central Tuamotas of French Polynesia, an example of Stage III in the evolution of an atoll. (With the permission of and copyrighted by Erwin Christian.)